THE
THEORETICAL
MINIMUM

THE THEORETICAL MINIMUM

WHAT YOU NEED *to* KNOW
to START DOING PHYSICS

LEONARD SUSSKIND

and

GEORGE HRABOVSKY

BASIC BOOKS
A Member of the Perseus Books Group
New York

Books published by Basic Books are available at special discounts for bulk purchases in the United States by corporations, institutions, and other organizations. For more information, please contact the Special Markets Department at the Perseus Books Group, 2300 Chestnut Street, Suite 200, Philadelphia, PA 19103, or call (800) 810-4145, ext. 5000, or e-mail special.markets@perseusbooks.com.

LCCN: 2012953679

ISBN 978-0-465-02811-5 (hardcover)
ISBN 978-0-465-03174-0 (e-book)

10 9 8 7 6 5 4 3 2 1

To our spouses—
those who have chosen to put up with us,
and to the students of Professor Susskind's
Continuing Education Courses

CONTENTS

Preface

I've always enjoyed explaining physics. For me it's much more than teaching: It's a way of thinking. Even when I'm at my desk doing research, there's a dialog going on in my head. Figuring out the best way to explain something is almost always the best way to understand it yourself.

About ten years ago someone asked me if I would teach a course for the public. As it happens, the Stanford area has a lot of people who once wanted to study physics, but life got in the way. They had had all kinds of careers but never forgot their one-time infatuation with the laws of the universe. Now, after a career or two, they wanted to get back into it, at least at a casual level.

Unfortunately there was not much opportunity for such folks to take courses. As a rule, Stanford and other universities don't allow outsiders into classes, and, for most of these grown-ups, going back to school as a full-time student is not a realistic option. That bothered me. There ought to be a way for people to develop their interest by interacting with active scientists, but there didn't seem to be one.

That's when I first found out about Stanford's Continuing Studies program. This program offers courses for people in the local nonacademic community. So I thought that it might just serve my purposes in finding someone to explain physics to, as well as their purposes, and it might also be fun to teach a course on modern physics. For one academic quarter anyhow.

It was fun. And it was very satisfying in a way that teaching undergraduate and graduate students was sometimes not. These students were there for only one reason: Not to get credit, not to get a degree, and not to be tested, but just to learn and indulge their curiosity. Also, having been "around the block" a few times, they were not at all afraid to ask questions, so the class had a lively vibrancy that academic classes often lack. I decided to do it again. And again.

What became clear after a couple of quarters is that the students were not completely satisfied with the layperson's courses I was teaching. They wanted more than the *Scientific American* experience. A lot of them had a bit of background, a bit of physics, a rusty but not dead knowledge of calculus, and some experience at solving technical problems. They were ready to try their hand at learning the real thing—with equations. The result was a sequence of courses intended to bring these students to the forefront of modern physics and cosmology.

Fortunately, someone (not I) had the bright idea to video-record the classes. They are out on the Internet, and it seems that they are tremendously popular: Stanford is not the only place with people hungry to learn physics. From all over the world I get thousands of e-mail messages. One of the main inquiries is whether I will ever convert the lectures into books? *The Theoretical Minimum* is the answer.

The term *theoretical minimum* was not my own invention. It originated with the great Russian physicist Lev Landau. The TM in Russia meant everything a student needed to know to work under Landau himself. Landau was a very demanding man: His theoretical minimum meant just about everything he knew, which of course no one else could possibly know.

I use the term differently. For me, the theoretical minimum means just what you need to know in order to proceed to the next level. It means not fat encyclopedic textbooks that explain everything, but thin books that explain everything important. The books closely follow the Internet courses that you will find on the Web.

Welcome, then, to *The Theoretical Miniumum*—Classical Mechanics, and good luck!

Leonard Susskind

Stanford, California, July 2012

I started to teach myself math and physics when I was eleven. That was forty years ago. A lot of things have happened since then—I am one of those individuals who got sidetracked by life. Still, I have learned a lot of math and physics. Despite the fact that people pay me to do research for them, I never pursued a degree.

For me, this book began with an e-mail. After watching the lectures that form the basis for the book, I wrote an e-mail to Leonard Susskind asking if he wanted to turn the lectures into a book. One thing led to another, and here we are.

We could not fit everything we wanted into this book, or it wouldn't be *The Theoretical Minimum*—Classical Mechanics, it would be A-Big-Fat-Mechanics-Book. That is what the Internet is for: Taking up large quantities of bandwidth to display stuff that doesn't fit elsewhere! You can find extra material at the website www.madscitech.org/tm. This material will include answers to the problems, demonstrations, and additional material that we couldn't put in the book.

I hope you enjoy reading this book as much as we enjoyed writing it.

George Hrabovsky

Madison, Wisconsin, July 2012

Lecture 1: The Nature of Classical Physics

Somewhere in Steinbeck country two tired men sit down at the side of the road. Lenny combs his beard with his fingers and says, "Tell me about the laws of physics, George." George looks down for a moment, then peers at Lenny over the tops of his glasses. "Okay, Lenny, but just the minimum."

What Is Classical Physics?

The term *classical physics* refers to physics before the advent of quantum mechanics. Classical physics includes Newton's equations for the motion of particles, the Maxwell-Faraday theory of electromagnetic fields, and Einstein's general theory of relativity. But it is more than just specific theories of specific phenomena; it is a set of principles and rules—an underlying logic—that governs all phenomena for which quantum uncertainty is not important. Those general rules are called *classical mechanics*.

The job of classical mechanics is to predict the future. The great eighteenth-century physicist Pierre-Simon Laplace laid it out in a famous quote:

> *We may regard the present state of the universe as the effect of its past and the cause of its future. An intellect which at a certain moment would know all forces that set nature in motion, and all positions of all items of which nature is composed, if this intellect were also vast enough to submit these data to analysis, it would embrace in a single formula the movements of the greatest bodies of the universe and those*

of the tiniest atom; for such an intellect nothing would be uncertain and the future just like the past would be present before its eyes.

In classical physics, if you know everything about a system at some instant of time, and you also know the equations that govern how the system changes, then you can predict the future. That's what we mean when we say that the classical laws of physics are *deterministic*. If we can say the same thing, but with the past and future reversed, then the same equations tell you everything about the past. Such a system is called *reversible*.

Simple Dynamical Systems and the Space of States

A collection of objects—particles, fields, waves, or whatever—is called a *system*. A system that is either the entire universe or is so isolated from everything else that it behaves as if nothing else exists is a *closed* system.

> **Exercise 1:** Since the notion is so important to theoretical physics, think about what a closed system is and speculate on whether closed systems can actually exist. What assumptions are implicit in establishing a closed system? What is an open system?

To get an idea of what deterministic and reversible mean, we are going to begin with some extremely simple closed systems. They are much simpler than the things we usually study in physics, but they satisfy rules that are rudimentary versions of the laws of classical mechanics. We begin with an example that is so simple it is trivial. Imagine an abstract object that has only one state. We could think of it as a coin glued to the table—forever showing heads. In physics jargon, the collection of all states occupied by a system is its space of states, or, more simply, its *state-space*. The state-space is not ordinary space; it's a mathematical set whose elements label the possible states of the system. Here the state-space consists of a single point—namely Heads (or just H)—because the system has only one state. Predicting the future of this system is extremely simple: Nothing ever happens and the outcome of any observation is always H.

The next simplest system has a state-space consisting of two points; in this case we have one abstract object and two possible states. Imagine a coin that can be either Heads or Tails (H or T). See Figure 1.

Figure 1: The space of two states.

In classical mechanics we assume that systems evolve smoothly, without any jumps or interruptions. Such behavior is said to be *continuous*. Obviously you cannot move between Heads and Tails smoothly. Moving, in this case, necessarily occurs in discrete jumps. So let's assume that time comes in discrete steps labeled by integers. A world whose evolution is discrete could be called *stroboscopic*.

A system that changes with time is called a *dynamical system*. A dynamical system consists of more than a space of states. It also entails a *law of motion*, or *dynamical law*. The dynamical law is a rule that tells us the next state given the current state.

One very simple dynamical law is that whatever the state at some instant, the next state is the same. In the case of our example, it has two possible histories: H H H H H H . . . and T T T T T . . . .

Another dynamical law dictates that whatever the current state, the next state is the opposite. We can make diagrams to illustrate these two laws. Figure 2 illustrates the first law, where the arrow from H goes to H and the arrow from T goes to T. Once again it is easy to predict the future: If you start with H, the system stays H; if you start with T, the system stays T.

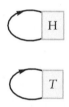

Figure 2: A dynamical law for a two-state system.

A diagram for the second possible law is shown in Figure 3, where the arrows lead from H to T and from T to H. You can still predict the future. For example, if you start with H the history will be H T H T H T H T H T . . . . If you start with T the history is T H T H T H T H . . . .

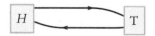

Figure 3: Another dynamical law for a two-state system.

We can even write these dynamical laws in equation form. The variables describing a system are called its *degrees of freedom.*

Our coin has one degree of freedom, which we can denote by the greek letter sigma, σ. Sigma has only two possible values; $\sigma = 1$ and $\sigma = -1$, respectively, for H and T. We also use a symbol to keep track of the time. When we are considering a continuous evolution in time, we can symbolize it with t. Here we have a discrete evolution and will use n. The state at time n is described by the symbol $\sigma(n)$, which stands for σ at n.

Let's write equations of evolution for the two laws. The first law says that no change takes place. In equation form,

$$\sigma(n+1) = \sigma(n).$$

In other words, whatever the value of σ at the nth step, it will have the same value at the next step.

The second equation of evolution has the form

$$\sigma(n+1) = -\sigma(n),$$

implying that the state flips during each step.

Because in each case the future behavior is completely determined by the initial state, such laws are deterministic. All the basic laws of classical mechanics are deterministic.

To make things more interesting, let's generalize the system by increasing the number of states. Instead of a coin, we could use a six-sided die, where we have six possible states (see Figure 4).

Now there are a great many possible laws, and they are not so easy to describe in words—or even in equations. The simplest way is to stick to diagrams such as Figure 5. Figure 5 says that given the numerical state of the die at time n, we increase the state one unit at the next instant $n+1$. That works fine until we get to 6, at which point the diagram tells you to go back to 1 and repeat the pattern. Such a pattern that is repeated endlessly is called a *cycle*. For example, if we start with 3 then the history is 3, 4, 5, 6, 1, 2, 3, 4, 5, 6, 1, 2, ... We'll call this pattern Dynamical Law 1.

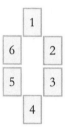

Figure 4: A six-state system.

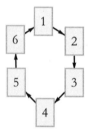

Figure 5: Dynamical Law 1.

Figure 6 shows another law, Dynamical Law 2. It looks a little messier than the last case, but it's logically identical—in each case the system endlessly cycles through the six possibilities. If we relabel the states, Dynamical Law 2 becomes identical to Dynamical Law 1.

Not all laws are logically the same. Consider, for example, the law shown in Figure 7. Dynamical Law 3 has two cycles. If you start on one of them, you can't get to the other. Nevertheless, this law is completely deterministic. Wherever you start, the future is determined. For example, if you start at 2, the history will be 2, 6, 1, 2, 6, 1, . . . and you will never get to 5. If you start at 5 the history is 5, 3, 4, 5, 3, 4, . . . and you will never get to 6.

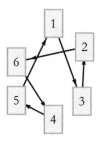

Figure 6: Dynamical Law 2.

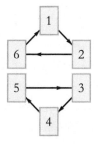

Figure 7: Dynamical Law 3.

Figure 8 shows Dynamical Law 4 with three cycles.

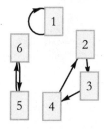

Figure 8: Dynamical Law 4.

It would take a long time to write out all of the possible dynamical laws for a six-state system.

> **Exercise 2: Can you think of a general way to classify the laws that are possible for a six-state system?**

Rules That Are Not Allowed: The Minus-First Law

According to the rules of classical physics, not all laws are legal. It's not enough for a dynamical law to be deterministic; it must also be reversible.

The meaning of *reversible*—in the context of physics—can be described a few different ways. The most concise description is to say that if you reverse all the arrows, the resulting law is still deterministic. Another way, is to say *the laws are deterministic into the past as well as the future*. Recall Laplace's remark, "for such an intellect nothing would be uncertain and the future just like the past would be present before its eyes." Can one conceive of laws that are deterministic into the future, but not into the past? In other words, can we formulate irreversible laws? Indeed we can. Consider Figure 9.

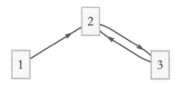

Figure 9: A system that is irreversible.

The law of Figure 9 does tell you, wherever you are, where to go next. If you are at 1, go to 2. If at 2, go to 3. If at 3, go to 2.

There is no ambiguity about the future. But the past is a different matter. Suppose you are at 2. Where were you just before that? You could have come from 3 or from 1. The diagram just does not tell you. Even worse, in terms of reversibility, there is no state that leads to 1; state 1 has no past. The law of Figure 9 is *irreversible*. It illustrates just the kind of situation that is prohibited by the principles of classical physics.

Notice that if you reverse the arrows in Figure 9 to give Figure 10, the corresponding law fails to tell you where to go in the future.

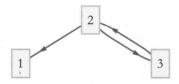

Figure 10: A system that is not deterministic into the future.

There is a very simple rule to tell when a diagram represents a deterministic reversible law. If every state has a single unique arrow leading into it, and a single arrow leading out of it, then it is a legal deterministic reversible law. Here is a slogan: *There must be one arrow to tell you where you're going and one to tell you where you came from.*

The rule that dynamical laws must be deterministic and reversible is so central to classical physics that we sometimes forget to mention it when teaching the subject. In fact, it doesn't even have a name. We could call it the first law, but unfortunately there are already two first laws—Newton's and the first law of thermodynamics. There is evan a zeroth law of thermodynamics. So we have to go back to a *minus-first law* to gain priority for what is undoubtedly the most fundamental of all physical laws—*the conservation of information*. The conservation of information is simply the rule that every state has one arrow in and one arrow out. It ensures that you never lose track of where you started.

The conservation of information is not a conventional conservation law. We will return to conservation laws after a digression into systems with infinitely many states.

Dynamical Systems with an Infinite Number of States

So far, all our examples have had state-spaces with only a finite number of states. There is no reason why you can't have a dynamical system with an infinite number of states. For example, imagine a line with an infinite number of discrete points along it—like a train track with an infinite sequence of stations in both directions. Suppose that a marker of some sort can jump from one point to another according to some rule. To describe such a system, we can label the points along the line by integers the same way we labeled the discrete instants of time above. Because we have already used the notation n for the discrete time steps, let's use an uppercase N for points on the track. A history of the marker would consist of a function $N(n)$, telling you the place along the track N at every time n. A short portion of this state-space is shown in Figure 11.

Figure 11: State-space for an infinite system.

A very simple dynamical law for such a system, shown in Figure 12, is to shift the marker one unit in the positive direction at each time step.

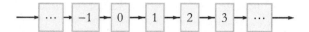

Figure 12: A dynamical rule for an infinite system.

This is allowable because each state has one arrow in and one arrow out. We can easily express this rule in the form of an equation.

$$N(n+1) = N(n) + 1 \tag{1}$$

Here are some other possible rules, but not all are allowable.

$$N(n+1) = N(n) - 1 \tag{2}$$

$$N(n+1) = N(n) + 2 \tag{3}$$

$$N(n+1) = N(n)^2 \tag{4}$$

$$N(n+1) = -1^{N(n)} N(n) \tag{5}$$

Exercise 3: Determine which of the dynamical laws shown in Eq.s (2) through (5) are allowable.

In Eq. (1), wherever you start, you will eventually get to every other point by either going to the future or going to the past. We say that there is a single infinite cycle. With Eq. (3), on the other hand, if you start at an odd value of N, you will never get to an even value, and vice versa. Thus we say there are two infinite cycles.

We can also add qualitatively different states to the system to create more cycles, as shown in Figure 13.

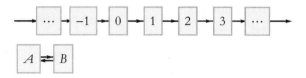

Figure 13: Breaking an infinite configuration space into
finite and infinite cycles.

If we start with a number, then we just keep proceeding through
the upper line, as in Figure 12. On the other hand, if we start at
A or *B*, then we cycle between them. Thus we can have mixtures
where we cycle around in some states, while in others we move
off to infinity.

Cycles and Conservation Laws

When the state-space is separated into several cycles, the system
remains in whatever cycle it started in. Each cycle has its own
dynamical rule, but they are all part of the same state-space
because they describe the same dynamical system. Let's consider
a system with three cycles. Each of states 1 and 2 belongs to its
own cycle, while 3 and 4 belong to the third (see Figure 14).

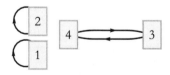

Figure 14: Separating the state-space into cycles.

Whenever a dynamical law divides the state-space into
such separate cycles, there is a memory of which cycle they
started in. Such a memory is called a *conservation law*; it tells us that

something is kept intact for all time. To make the conservation law quantitative, we give each cycle a numerical value called Q. In the example in Figure 15 the three cycles are labeled $Q = +1$, $Q = -1$, and $Q = 0$. Whatever the value of Q, it remains the same for all time because the dynamical law does not allow jumping from one cycle to another. Simply stated, Q is conserved.

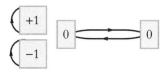

Figure 15: Labeling the cycles with specific values of a conserved quantity.

In later chapters we will take up the problem of continuous motion in which both time and the state-space are continuous. All of the things that we discussed for simple discrete systems have their analogs for the more realistic systems but it will take several chapters before we see how they all play out.

The Limits of Precision

Laplace may have been overly optimistic about how predictable the world is, even in classical physics. He certainly would have agreed that predicting the future would require a perfect knowledge of the dynamical laws governing the world, as well as tremendous computing power—what he called an "intellect vast enough to submit these data to analysis." But there is another element that he may have underestimated: the ability to know the initial conditions with almost perfect precision. Imagine a die with a million faces, each of which is labeled with a symbol similar in appearance to the usual single-digit integers, but with enough slight differences so that there are a million distinguishable labels. If one knew the dynamical law, and if one were able to recognize the initial label, one could predict the future history of the die. However, if Laplace's vast intellect suffered from a slight vision impairment, so that he was unable to distinguish among similar labels, his predicting ability would be limited.

In the real world, it's even worse; the space of states is not only huge in its number of points—it is continuously infinite. In other words, it is labeled by a collection of real numbers such as the coordinates of the particles. Real numbers are so dense that every one of them is arbitrarily close in value to an infinite number of neighbors. The ability to distinguish the neighboring values of these numbers is the "resolving power" of any experiment, and for any real observer it is limited. In principle we cannot know the initial conditions with infinite precision. In most cases the tiniest differences in the initial conditions—the starting state—leads to large eventual differences in outcomes. This phenomenon is called *chaos*. If a system is chaotic (most are), then it implies that however good the resolving power may be, the time over which the system is predictable is limited. Perfect predictability is not achievable, simply because we are limited in our resolving power.

Interlude 1: Spaces, Trigonometry, and Vectors

"Where are we, George?"

George pulled out his map and spread it out in front of Lenny. "We're right here Lenny, coordinates 36.60709N, –121.618652W."

"Huh? What's a coordinate George?"

Coordinates

To describe points quantitatively, we need to have a coordinate system. Constructing a coordinate system begins with choosing a point of space to be the *origin*. Sometimes the origin is chosen to make the equations especially simple. For example, the theory of the solar system would look more complicated if we put the origin anywhere but at the Sun. Strictly speaking, the location of the origin is arbitrary—put it anywhere—but once it is chosen, stick with the choice.

The next step is to choose three perpendicular axes. Again, their location is somewhat arbitrary as long as they are perpendicular. The axes are usually called x, y, and z but we can also call them x_1, x_2, and x_3. Such a system of axes is called a *Cartesian coordinate system*, as in Figure 1.

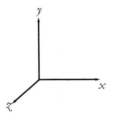

Figure 1. A three-dimensional Cartesian coordinate system.

We want to describe a certain point in space; call it $\mathcal{P}$. It can be located by giving the x, y, z coordinates of the point. In other words, we identify the point $\mathcal{P}$ with the ordered triple of numbers (x, y, z) (see Figure 2).

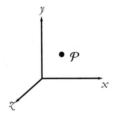

Figure 2. A point in Cartesian space.

The x coordinate represents the perpendicular distance of $\mathcal{P}$ from the plane defined by setting $x = 0$ (see Figure 3). The same is true for the y and z coordinates. Because the coordinates represent distances they are measured in units of length, such as meters.

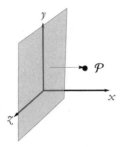

Figure 3: A plane defined by setting $x = 0$, and the distance
to $\mathcal{P}$ along the x axis.

When we study motion, we also need to keep track of time. Again we start with an origin—that is, the zero of time. We could pick the Big Bang to be the origin, or the birth of Jesus, or just the start of an experiment. But once we pick it, we don't change it.

Next we need to fix a direction of time. The usual convention is that positive times are to the future of the origin and negative times are to the past. We could do it the other way, but we won't.

Finally, we need units for time. Seconds are the physicist's customary units, but hours, nanoseconds, or years are also possible. Once having picked the units and the origin, we can label any time by a number t.

There are two implicit assumptions about time in classical mechanics. The first is that time runs uniformly—an interval of 1 second has exactly the same meaning at one time as at another. For example, it took the same number of seconds for a weight to fall from the Tower of Pisa in Galileo's time as it takes in our time. One second meant the same thing then as it does now.

The other assumption is that times can be compared at

different locations. This means that clocks located in different places can be synchronized. Given these assumptions, the four coordinates—*x*, *y*, *z*, *t*—define a *reference frame*. Any event in the reference frame must be assigned a value for each of the coordinates.

Given the function $f(t) = t^2$, we can plot the points on a coordinate system. We will use one axis for time, *t*, and another for the function, $f(t)$ (see Figure 4).

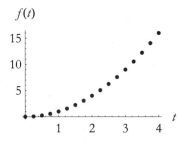

Figure 4: Plotting the points of $f(t) = t^2$.

We can also connect the dots with curves to fill in the spaces between the points (see Figure 5).

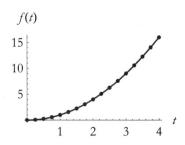

Figure 5: Joining the plotted points with curves.

In this way we can visualize functions.

Exercise 1: Using a graphing calculator or a program like *Mathematica*, plot each of the following functions. See the next section if you are unfamiliar with the trigonometric functions.

$f(t) = t^4 + 3t^3 - 12t^2 + t - 6$
$g(x) = \sin x - \cos x$
$\theta(\alpha) = e^\alpha + \alpha \ln \alpha$
$x(t) = \sin^2 x - \cos x$

Trigonometry

If you have not studied trigonometry, or if you studied it a long time ago, then this section is for you.

We use trigonometry in physics all the time; it is everywhere. So you need to be familiar with some of the ideas, symbols, and methods used in trigonometry. To begin with, in physics we do not generally use the degree as a measure of angle. Instead we use the *radian*; we say that there are 2π radians in 360°, or 1 radian $= \pi/180°$, thus $90° = \pi/2$ radians, and $30° = \pi/6$ radians. Thus a radian is about 57° (see Figure 6).

The trigonometric functions are defined in terms of properties of right triangles. Figure 7 illustrates the right triangle and its hypotenuse c, base b, and altitude a. The greek letter theta, θ, is defined to be the angle opposite the altitude, and the greek letter phi, ϕ, is defined to be the angle opposite the base.

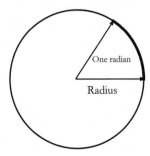

Figure 6: The radian as the angle subtended by an arc equal to the radius of the circle.

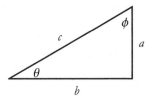

Figure 7: A right triangle with segments and angles indicated.

We define the functions sine (sin), cosine (cos), and tangent (tan), as ratios of the various sides according to the following relationships:

$$\sin \theta = \frac{a}{c}$$

$$\cos \theta = \frac{b}{c}$$

$$\tan \theta = \frac{a}{b} = \frac{\sin \theta}{\cos \theta}.$$

We can graph these functions to see how they vary (see Figures 8 through 10).

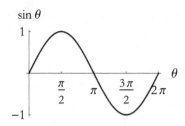

Figure 8: Graph of the sine function.

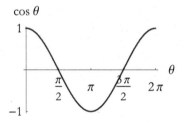

Figure 9: Graph of the cosine function.

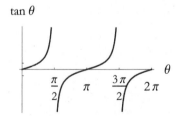

Figure 10: Graph of the tangent function.

There are a couple of useful things to know about the trigonometric functions. The first is that we can draw a triangle within a circle, with the center of the circle located at the origin of a Cartesian coordinate system, as in Figure 11.

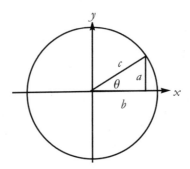

Figure 11: A right triangle drawn in a circle.

Here the line connecting the center of the circle to any point along its circumference forms the hypotenuse of a right triangle, and the horizontal and vertical components of the point are the base and altitude of that triangle. The position of a point can be specified by two coordinates, x and y, where

$$x = c \cos \theta$$

and

$$y = c \sin \theta.$$

This is a very useful relationship between right triangles and circles.

Suppose a certain angle θ is the sum or difference of two other angles using the greek letters alpha, α, and beta, β, we can write this angle, θ, as $\alpha \pm \beta$. The trigonometric functions of $\alpha \pm \beta$ can be expressed in terms of the trigonometric functions of α and β.

$$\sin (\alpha + \beta) = \sin \alpha \cos \beta + \cos \alpha \sin \beta$$
$$\sin (\alpha - \beta) = \sin \alpha \cos \beta - \cos \alpha \sin \beta$$
$$\cos (\alpha + \beta) = \cos \alpha \cos \beta - \sin \alpha \sin \beta$$
$$\cos (\alpha - \beta) = \cos \alpha \cos \beta + \sin \alpha \sin \beta.$$

A final—very useful—identity is

$$\sin^2 \theta + \cos^2 \theta = 1. \tag{1}$$

(Notice the notation used here: $\sin^2 \theta = \sin \theta \sin \theta$.) This equation is the Pythagorean theorem in disguise. If we choose the radius of the circle in Figure 11 to be 1, then the sides a and b are the sine and cosine of θ, and the hypotenuse is 1. Equation (1) is the familiar relation among the three sides of a right triangle: $a^2 + b^2 = c^2$.

Vectors

Vector notation is another mathematical subject that we assume you have seen before, but—just to level the playing field—let's review vector methods in ordinary three-dimensional space.

A *vector* can be thought of as an object that has both a length (or *magnitude*) and a direction in space. An example is displacement. If an object is moved from some particular starting location, it is not enough to know how far it is moved in order to know where it winds up. One also has to know the direction of the displacement. Displacement is the simplest example of a vector quantity. Graphically, a vector is depicted as an arrow with a length and direction, as shown in Figure 12.

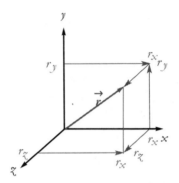

Figure 12: A vector $\vec{r}$ in Cartesian coordinates.

Symbolically vectors are represented by placing arrows over them. Thus the symbol for displacement is $\vec{r}$. The magnitude, or length, of a vector is expressed in absolute-value notation. Thus the length of $\vec{r}$ is denoted $\left|\vec{r}\right|$.

Here are some operations that can be done with vectors. First of all, you can multiply them by ordinary real numbers. When dealing with vectors you will often see such real numbers given the special name *scalar*. Multiplying by a positive number just multiplies the length of the vector by that number. But you can also multiply by a negative number, which reverses the direction of the vector. For example $-2\,\vec{r}$ is the vector that is twice as long as $\vec{r}$ but points in the opposite direction.

Vectors may be added. To add $\vec{A}$ and $\vec{B}$, place them as shown in Figure 13 to form a quadrilateral (this way the directions of the vectors are preserved). The sum of the vectors is the length and angle of the diagonal.

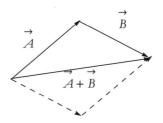

Figure 13: Adding vectors.

If vectors can be added and if they can be multiplied by negative numbers then they can be subtracted.

Exercise 2: Work out the rule for vector subtraction.

Vectors can also be described in component form. We begin with three perpendicular axes x, y, z. Next, we define three *unit vectors* that lie along these axes and have unit length. The unit vectors along the coordinate axes are called *basis vectors*. The three basis vectors for Cartesian coordinates are traditionally called $\hat{i}$, $\hat{j}$, and $\hat{k}$ (see Figure 14). More generally, we write $\hat{e}_1$, $\hat{e}_2$, and $\hat{e}_3$ when we refer to (x_1, x_2, x_3), where the symbol ^ (known as a carat) tells us we are dealing with unit (or basis) vectors. The basis vectors are useful because any vector $\vec{V}$ can be written in terms of them in the following way:

$$\vec{V} = V_x\,\hat{i} + V_y\,\hat{j} + V_z\,\hat{k}. \qquad (2)$$

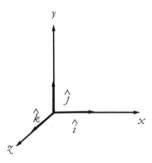

Figure 14: Basis vectors for a Cartesian coordinate system.

The quantities V_x, V_y, and V_z are numerical coefficients that are needed to add up the basis vectors to give $\vec{V}$. They are also called the *components* of $\vec{V}$. We can say that Eq. (2) is a *linear combination* of basis vectors. This is a fancy way of saying that we add the basis vectors along with any relevant factors. Vector components can be positive or negative. We can also write a vector as a list of its components—in this case $\left(V_x, V_y, V_z\right)$. The magnitude of a vector can be given in terms of its components by applying the three-dimensional Pythagorean theorem.

$$\left|\vec{V}\right| = \sqrt{V_x^2 + V_y^2 + V_z^2} \qquad (3)$$

We can multiply a vector $\vec{V}$ by a scalar, α, in terms of components by multiplying each component by α.

$$\alpha \vec{V} = \left(\alpha V_x, \alpha V_y, \alpha V_z\right)$$

We can write the sum of two vectors as the sum of the

corresponding components.

$$\left(\vec{A} + \vec{B}\right)_x = \left(A_x + B_x\right)$$

$$\left(\vec{A} + \vec{B}\right)_y = \left(A_y + B_y\right)$$

$$\left(\vec{A} + \vec{B}\right)_z = \left(A_z + B_z\right).$$

Can we multiply vectors? Yes, and there is more than one way. One type of product—the cross product—gives another vector. For now, we will not worry about the cross product and only consider the other method, the *dot product*. The dot product of two vectors is an ordinary number, a scalar. For vectors $\vec{A}$ and $\vec{B}$ it is defined as follows:

$$\vec{A} \cdot \vec{B} = \left|\vec{A}\right|\left|\vec{B}\right| \cos \theta.$$

Here θ is the angle between the vectors. In ordinary language, the dot product is the product of the magnitudes of the two vectors and the cosine of the angle between them.

The dot product can also be defined in terms of components in the form

$$\vec{A} \cdot \vec{B} = A_x B_x + A_y B_y + A_z B_z.$$

This makes it easy to compute dot products given the components of the vectors.

Exercise 3: Show that the magnitude of a vector satisfies
$$\left|\vec{A}\right|^2 = \vec{A} \cdot \vec{A}.$$

Exercise 4: Let $\left(A_x = 2, \quad A_y = -3, A_z = 1\right)$ **and** $\left(B_x = -4, B_y = -3, B_z = 2\right)$. **Compute the mag- nitude of a** $\vec{A}$ **and** $\vec{B}$, **their dot product, and the angle between them.**

An important property of the dot product is that it is zero if the vectors are *orthogonal* (perpendicular). Keep this in mind because we will have occasion to use it to show that vectors are orthogonal.

Exercise 5: Determine which pair of vectors are orthogonal. (1, 1, 1) (2, -1, 3) (3, 1, 0) (-3, 0, 2)

Exercise 6: Can you explain why the dot product of two vectors that are orthogonal is 0?

Lecture 2: Motion

Lenny complained, "George, this jumpy stroboscopic stuff makes me nervous. Is time really so bumpy? I wish things would go a little more smoothly."

George thought for a moment, wiping the blackboard. "Okay, Lenny, today let's study systems that do change smoothly."

Mathematical Interlude: Differential Calculus

In this book we will mostly be dealing with how various quantities change with time. Most of classical mechanics deals with things that change smoothly—*continuously* is the mathematical term—as time changes continuously. Dynamical laws that update a state will have to involve such continuous changes of time, unlike the stroboscopic changes of the first lecture. Thus we will be interested in functions of the independent variable t.

To cope, mathematically, with continuous changes, we use the mathematics of calculus. Calculus is about limits, so let's get that idea in place. Suppose we have a sequence of numbers, l_1, l_2, l_3, ..., that get closer and closer to some value L. Here is an example: 0.9, 0.99, 0.999, 0.9999, ... The limit of this sequence is 1. None of the entries is equal to 1, but they get closer and closer to that value. To indicate this we write

$$\lim_{i \to \infty} l_i = L.$$

In words, L is the limit of l_i as i goes to infinity.

We can apply the same idea to functions. Suppose we have a function, $f(t)$, and we want to describe how it varies as t gets closer and closer to some value, say a. If $f(t)$ gets arbitrarily close to L as t tends to a, then we say that the limit of $f(t)$ as t approaches a is the number L. Symbolically,

$$\lim_{t \to a} f(t) = L.$$

Let $f(t)$ be a function of the variable t. As t varies, so will $f(t)$. Differential calculus deals with the rate of change of such functions. The idea is to start with $f(t)$ at some instant, and then to change the time by a little bit and see how much $f(t)$ changes. The rate of change is defined as the ratio of the change in f to the change in t. We denote the change in a quantity with the uppercase greek letter delta, Δ. Let the change in t be called Δt. (This is not $\Delta \times t$, this is a change in t.) Over the interval Δt, f changes from $f(t)$ to $f(t + \Delta t)$. The change in f, denoted Δf, is then given by

$$\Delta f = f(t + \Delta t) - f(t).$$

To define the rate of change precisely at time t, we must let Δt shrink to zero. Of course, when we do that Δf also shrinks to zero, but if we divide Δf by Δt, the ratio will tend to a limit. That limit is the derivative of $f(t)$ with respect to t,

$$\frac{df(t)}{dt} = \lim_{\Delta t \to 0} \frac{\Delta f}{\Delta t} = \lim_{\Delta t \to 0} \frac{f(t + \Delta t) - f(t)}{\Delta t}. \tag{1}$$

A rigorous mathematician might frown on the idea that $\frac{df(t)}{dt}$ is the ratio of two differentials, but you will rarely make a mistake this way.

Let's calculate a few derivatives. Begin with functions defined by powers of t. In particular, let's illustrate the method by calculating the derivative of $f(t) = t^2$. We apply Eq. (1) and begin by defining $f(t + \Delta t)$:

$$f(t + \Delta t) = (t + \Delta t)^2.$$

We can calculate $(t + \Delta t)^2$ by direct multiplication or we can use the binomial theorem. Either way,

$$f(t + \Delta t) = t^2 + 2t\Delta t + \Delta t^2.$$

We now subtract $f(t)$:

$$
\begin{aligned}
f(t + \Delta t) - f(t) &= t^2 + 2t\Delta t + \Delta t^2 - t^2 \\
&= 2t\Delta t + \Delta t^2.
\end{aligned}
$$

The next step is to divide by Δt:

$$
\begin{aligned}
\frac{f(t + \Delta t) - f(t)}{\Delta t} &= \frac{2t\Delta t + \Delta t^2}{\Delta t} \\
&= 2t + \Delta t.
\end{aligned}
$$

Now it's easy to take the limit $\Delta t \to 0$. The first term does not depend on Δt and survives, but the second term tends to zero and just disappears. This is something to keep in mind: Terms of higher order in Δt can be ignored when you calculate derivatives. Thus

$$\lim_{\Delta t \to 0} \frac{f(t + \Delta t) - f(t)}{\Delta t} = 2t$$

So the derivative of t^2 is

$$\frac{d\left(t^2\right)}{dt} = 2t$$

Next let us consider a general power, $f(t) = t^n$. To caclulate its derivative, we have to calculate $f(t + \Delta t) = (t + \Delta t)^n$. Here, high school algebra comes in handy: The result is given by the binomial theorem. Given two numbers, a and b, we would like to calculate $(a + b)^n$. The binomial theorem gives

$$(a + b)^n = a^n + na^{n-1}b + \frac{n(n - 1)}{2}a^{n-2}b^2 +$$
$$\frac{n(n - 1)(n - 2)}{3}a^{n-3}b^3 +$$
$$\cdots + b^n$$

How long does the expression go on? If n is an integer, it eventually terminates after $n + 1$ terms. But the binomial theorem is more general than that; in fact, n can be any real or complex number. If n is not an integer, however, the expression never terminates; it is an infinite series. Happily, for our purposes, only the first two terms are important.

To calculate $(t + \Delta t)^n$, all we have to do is plug in $a = t$ and $b = \Delta t$ to get

$$f(t + \Delta t) = (t + \Delta t)^n$$
$$= t^n + nt^{n-1}\Delta t + \cdots .$$

All the terms represented by the dots shrink to zero in the limit, so we ignore them.

Now subtract $f(t)$ (or t^n),

$$\Delta f = f(t + \Delta t) - f(t)$$
$$= t^n + nt^{n-1}\Delta t +$$

$$\frac{n(n-1)}{2}t^{n-2}\Delta t^2 + \cdots - t^n$$

$$= nt^{n-1}\Delta t +$$

$$\frac{n(n-1)}{2}t^{n-2}\Delta t^2 + \cdots.$$

Now divide by Δt,

$$\frac{\Delta f}{\Delta t} = nt^{n-1} + \frac{n(n-1)}{2}t^{n-2}\Delta t + \cdots.$$

and let $\Delta t \to 0$. The derivative is then

$$\frac{d(t^n)}{dt} = nt^{n-1}.$$

One important point is that this relation holds even if n is not an integer; n can be any real or complex number.

Here are some special cases of derivatives: If $n = 0$, then $f(t)$ is just the number 1. The derivative is zero—this is the case for any function that does't change. If $n = 1$, then $f(t) = t$ and the derivative is 1—this is always true when you take the derivative of something with respect to itself. Here are some derivatives of powers

$$\frac{d(t^2)}{dt} = 2t$$

$$\frac{d(t^3)}{dt} = 3t^2$$

$$\frac{d(t^4)}{dt} = 4t^3$$

$$\frac{d(t^n)}{dt} = nt^{n-1}.$$

For future reference, here are some other derivatives:

$$\frac{d(\sin t)}{dt} = \cos t$$

$$\frac{d(\cos t)}{dt} = -\sin t$$

$$\frac{d(e^t)}{dt} = e^t \qquad (2)$$

$$\frac{d(\log t)}{dt} = \frac{1}{t}.$$

One comment about the third formula in Eq. (2), $\frac{d(e^t)}{dt} = e^t$. The meaning of e^t is pretty clear if t is an integer. For example, $e^3 = e \times e \times e$. Its meaning for non-integers is not obvious. Basically, e^t is defined by the property that its derivative is equal to itself. So the third formula is really a definition.

There are a few useful rules to remember about derivatives. You can prove them all if you want a challenging exercise. The first is the fact that the derivative of a constant is always 0. This makes sense; the derivative is the rate of change, and a constant never changes, so

$$\frac{dc}{dt} = 0.$$

The derivative of a constant times a function is the constant times the derivative of the function:

$$\frac{d(cf)}{dt} = c\frac{df}{dt}.$$

Suppose we have two functions, $f(t)$ and $g(t)$. Their sum is also a function and its derivative is given by

$$\frac{d(f+g)}{dt} = \frac{d(f)}{dt} + \frac{d(g)}{dt}.$$

This is called the *sum rule*.

Their product of two functions is another function, and its derivative is

$$\frac{d(fg)}{dt} = f(t)\frac{d(g)}{dt} + g(t)\frac{d(f)}{dt}.$$

Not surprisingly, this is called the *product rule*.

Next, suppose that $g(t)$ is a function of t, and $f(g)$ is a function of g. That makes f an *implicit function* of t. If you want to know what f is for some t, you first compute $g(t)$. Then, knowing g, you compute $f(g)$. It's easy to calculate the t-derivative of f:

$$\frac{df}{dt} = \frac{df}{dg}\frac{dg}{dt}.$$

This is called the *chain rule*. This would obviously be true if the derivatives were really ratios; in that case, the dg's would cancel in the numerator and denominator. In fact, this is one of those cases where the naive answer is correct. The important thing to remember about using the chain rule is that you invent an intermediate function, $g(t)$, to simplify $f(t)$ making it $f(g)$. For example, if

$$f(t) = \ln t^3$$

and we need to find $\frac{df}{dt}$, then the t^3 inside the logarithm might be a problem. Therefore, we invent the intermediate function $g = t^3$, so we have $f(g) = \ln g$. We can then apply the chain rule.

$$\frac{df}{dt} = \frac{df}{dg}\frac{dg}{dt}.$$

We can use our differentiation formulas to note that $\frac{df}{dg} = \frac{1}{g}$ and $\frac{dg}{dt} = 3t^2$, so

$$\frac{df}{dt} = \frac{3t^2}{g}.$$

We can substitute $g = t^3$, to get

$$\frac{df}{dt} = \frac{3t^2}{t^3} = \frac{3}{t}.$$

That is how to use the chain rule.

Using these rules, you can calculate a lot of derivatives. That's basically all there is to differential calculus.

Exercise 1: Calculate the derivatives of each of these functions.

$$f(t) = t^4 + 3t^3 - 12t^2 + t - 6$$
$$g(x) = \sin x - \cos x$$
$$\theta(\alpha) = e^\alpha + \alpha \ln \alpha$$
$$x(t) = \sin^2 x - \cos x$$

Exercise 2: The derivative of a derivative is called the second derivative and is written $\dfrac{d^2 f(t)}{dt^2}$. Take the second derivative of each of the functions listed above.

Exercise 3: Use the chain rule to find the derivatives of each of the following functions.

$$g(t) = \sin(t^2) - \cos(t^2)$$
$$\theta(\alpha) = e^{3\alpha} + 3\alpha \ln(3\alpha)$$
$$x(t) = \sin^2(t^2) - \cos(t^2)$$

Exercise 4: Prove the sum rule (fairly easy), the product rule (easy if you know the trick), and the chain rule (fairly easy).

Exercise 5: Prove each of the formulas in Eq.s (2). *Hint: Look up trigonometric identities and limit properties in a reference book.*

Particle Motion

The concept of a point particle is an idealization. No object is so small that it is a point—not even an electron. But in many situations we can ignore the extended structure of objects and treat them as points. For example, the planet Earth is obviously not a point, but in calculating its orbit around the Sun, we can ignore the size of Earth to a high degree of accuracy.

The position of a particle is specified by giving a value for each of the three spatial coordinates, and the motion of the particle is defined by its position at every time. Mathematically, we can specify a position by giving the three spatial coordinates as functions of t: $x(t)$, $y(t)$, $z(t)$.

The position can also be thought of as a vector $\vec{r}(t)$ whose components are x, y, z at time t. The path of the particle—its *trajectory*—is specified by $\vec{r}(t)$. The job of classical mechanics is to figure out $\vec{r}(t)$ from some initial condition and some dynamical law.

Next to its position, the most important thing about a particle is its velocity. Velocity is also a vector. To define it we need some calculus. Here is how we do it:

Consider the displacement of the particle between time t and a little bit later at time $t + \Delta t$. During that time interval the particle moves from $x(t)$, $y(t)$, $z(t)$ to $x(t + \Delta t)$, $y(t + \Delta t)$, $z(t + \Delta t)$, or, in vector notation, from $\vec{r}(t)$ to $\vec{r}(t + \Delta t)$. The displacement is defined as

$$\Delta x = x(t + \Delta t) - x(t)$$
$$\Delta y = y(t + \Delta t) - y(t)$$
$$\Delta z = z(t + \Delta t) - z(t)$$

or

$$\Delta \vec{r} = \vec{r}(t + \Delta t) - \vec{r}(t).$$

The displacement is the small distance that the particle moves in the small time Δt. To get the velocity, we divide the displacement by Δt and take the limit as Δt shrinks to zero. For example,

$$v_x = \lim_{\Delta t \to 0} \frac{\Delta x}{\Delta t}.$$

This—of course—is the definition of the derivative of x with respect to t.

$$v_x = \frac{dx}{dt} = \dot{x}$$

$$v_y = \frac{dy}{dt} = \dot{y}$$

$$v_z = \frac{dz}{dt} = \dot{z}.$$

Placing a dot over a quantity is standard shorthand for taking the time derivative. This convention can be used to denote the time derivative of anything, not just the position of a particle. For example, if T stands for the temperature of a tub of hot water, then $\dot{T}$ will represent the rate of change of the temperature with time. It will be used over and over, so get familiar with it.

It gets tiresome to keep writing x, y, z, so we will often condense the notation. The three coordinates x, y, z are collectively denoted by x_i and the velocity components by v_i:

$$v_i = \frac{dx_i}{dt} = \dot{x}_i$$

where i takes the values x, y, z, or, in vector notation

$$\vec{v} = \frac{d\vec{r}}{dt} = \dot{\vec{r}}.$$

The velocity vector has a magnitude $\left|\vec{v}\right|$,

$$\left|\vec{v}\right|^2 = v_x{}^2 + v_y{}^2 + v_z{}^2,$$

this represents how fast the particle is moving, without regard to the direction. The magnitude $\left|\vec{v}\right|$ is called *speed*.

Acceleration is the quantity that tells you how the velocity is changing. If an object is moving with a constant velocity vector, it experiences no acceleration. A constant velocity vector implies not only a constant speed but also a constant direction. You feel acceleration only when your velocity vector changes, either in magnitude or direction. In fact, acceleration is the time derivative of velocity:

$$a_i = \frac{dv_i}{dt} = \dot{v}_i$$

or, in vector notation,

$$\vec{a} = \dot{\vec{v}}.$$

Because v_i is the time derivative of x_i and a_i is the time derivative of v_i, it follows that acceleration is the second time-derivative of x_i,

$$a_i = \frac{d^2 x_i}{dt^2} = \ddot{x}_i,$$

where the double-dot notation means the second time-derivative.

Examples of Motion

Suppose a particle starts to move at time $t = 0$ according to the equations

$$x(t) = 0$$
$$y(t) = 0$$
$$z(t) = z(0) + v(0)t - \frac{1}{2}gt^2$$

The particle evidently has no motion in the x and y directions but moves along the z axis. The constants $z(0)$ and $v(0)$ represent the initial values of the position and velocity along the z direction at $t = 0$. We also consider g to be a constant.

Let's calculate the velocity by differentiating with respect to time.

$$v_x(t) = 0$$
$$v_y(t) = 0$$
$$v_z(t) = v(0) - gt.$$

The x and y components of velocity are zero at all times. The z component of velocity starts out at $t = 0$ being equal to $v(0)$. In other words, $v(0)$ is the initial condition for velocity.

As time progresses, the $-gt$ term becomes nonzero. Eventually, it will overtake the initial value of the velocity, and the particle will be found moving along in the negative z direction.

Now let's calculate the acceleration by differentiating with respect to time again.

$$a_x(t) = 0$$
$$a_y(t) = 0$$
$$a_z(t) = -g.$$

The acceleration along the z axis is constant and negative. If the z axis were to represent altitude, the particle would accelerate downward in just the way a falling object would.

Next let's consider an oscillating particle that moves back and forth along the x axis. Because there is no motion in the other two directions, we will ignore them. A simple oscillatory motion uses trigonometric functions:

$$x(t) = \sin \omega t$$

where the lowercase greek letter omega, ω, is a constant. The larger ω, the more rapid the oscillation. This kind of motion is called *simple harmonic motion* (see Figure 1).

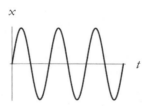

Figure 1: Simple harmonic motion.

Let's compute the velocity and acceleration. To do so, we need to differentiate $x(t)$ with respect to time. Here is the result of the first time-derivative:

$$v_x = \frac{d}{dt} \sin \omega t.$$

We have the sine of a product. We can relabel this product as $b = \omega t$:

$$v_x = \frac{d}{dt} \sin b.$$

Using the chain rule,

$$v_x = \frac{d}{db} \sin b \frac{d b}{dt}$$

or

$$v_x = \cos b \frac{d}{dt} (\omega t)$$

or

$$v_x = \omega \cos \omega t.$$

We get the acceleration by similar means:

$$a_x = -\omega^2 \sin \omega t.$$

Notice some interesting things. Whenever the position x is at its maximum or minimum, the velocity is zero. The opposite is also true: When the position is at $x = 0$, then velocity is either a maximum or a minimum. We say that position and velocity are $90°$ out of phase. You can see this in Figure 2, representing $x(t)$, and Figure 3, representing $v(t)$.

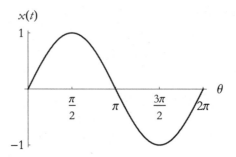

Figure 2: Representing position.

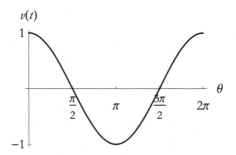

Figure 3: Representing velocity.

The position and acceleration are also related, both being proportional to $\sin \omega t$. But notice the minus sign in the acceleration. That minus sign says that whenever x is positive (negative), the acceleration is negative (positive). In other words, wherever the particle is, it is being accelerated back toward the origin. In technical terms, the position and acceleration are 180° out of phase.

Exercise 6: How long does it take for the oscillating particle to go through one full cycle of motion?

Next, let's consider a particle moving with uniform circular motion about the origin. This means that it is moving in a circle at a constant speed. For this purpose, we can ignore the z axis and think of the motion in the x, y plane. To describe it we must have two functions, $x(t)$ and $y(t)$. To be specific we will choose the particle to move in the counterclockwise direction. Let the radius of the orbit be R.

It is helpful to visualize the motion by projecting it onto the two axes. As the particle revolves around the origin, x oscillates between $x = -R$ and $x = R$. The same is true of the y coordinate. But the two coordinates are 90° out of phase; when x is maximum y is zero, and vice versa.

The most general (counterclockwise) uniform circular motion about the origin has the mathematical form

$$x(t) = R \cos \omega t$$
$$y(t) = R \sin \omega t.$$

Here the parameter ω is called the *angular frequency*. It is defined as the number of radians that the angle advances in unit time. It also has to do with how long it takes to go one full revolution, the period of motion—the same as we found in Exercise 6:

$$T = \frac{2\pi}{\omega}$$

Now it is easy to calculate the components of velocity and acceleration by differentiation:

$$v_x = -R \omega \sin \omega t$$
$$v_y = R \omega \cos \omega t$$
$$a_x = -R \omega^2 \cos \omega t \qquad (3)$$
$$a_y = -R \omega^2 \sin \omega t$$

This shows an interesting property of circular motion that Newton used in analyzing the motion of the moon: The acceleration of a circular orbit is parallel to the position vector, but it is oppositely directed. In other words, the acceleration vector points radially inward toward the origin.

Exercise 7: Show that the position and velocity vectors are orthogonal.

Exercise 8: Calculate the velocity, speed, and acceleration for each of the following position vectors. If you have graphing software, plot each position vector, each velocity vector, and each acceleration vector.

$$\vec{r} = (\cos \omega t, e^{\omega t})$$

$$\vec{r} = (\cos (\omega t - \phi), \sin (\omega t - \phi))$$

$$\vec{r} = \left(c \cos^3 t, c \sin^3 t\right)$$

$$\vec{r} = (c (t - \sin t), c (1 - \cos t))$$

Interlude 2: Integral Calculus

"George, I really like doing things backward. Can we do differentiation backward?"

"Sure we can, Lenny. It's called integration."

Integral Calculus

Differential calculus has to do with rates of change. Integral calculus has to do with sums of many tiny incremental quantities. It's not immediately obvious that these have anything to do with each other, but they do.

We begin with the graph of a function $f(t)$, as in Figure 1.

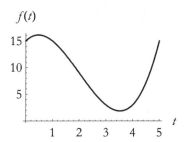

Figure 1: The behavior of $f(t)$.

The central problem of integral calculus is to calculate the area under the curve defined by $f(t)$. To make the problem well defined, we consider the function between two values that we call *limits of integration*, $t = a$ and $t = b$. The area we want to calculate is the area of the shaded region in Figure 2.

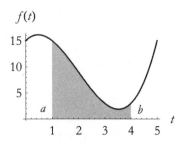

$f(t)$

Figure 2: The limits of integration.

In order to do this, we break the region into very thin rectangles and add their areas (see Figure 3).

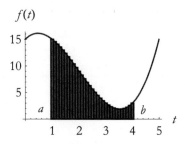

$f(t)$

Figure 3: An illustration of integration.

Of course this involves an approximation, but it becomes accurate if we let the width of the rectangles tend to zero. In order to carry out this procedure, we first divide the interval between $t = a$ and $t = b$ into a number, N, of subintervals—each of width Δt. Consider the rectangle located at a specific value of t. The width is Δt and the height is the local value of $f(t)$. It follows that the area of a single rectangle, δA, is

$$\delta A = f(t)\, \Delta t.$$

Now we add up all the areas of the individual rectangles to get an approximation to the area that we are seeking. The approximate answer is

$$A = \sum_i f(t_i) \Delta t$$

where the uppercase greek letter sigma, Σ, indicates a sum of successive values defined by i. So, if $N = 3$, then

$$A = \sum_i^3 f(t_i) \Delta t$$

$$= f(t_1) \Delta t + f(t_2) \Delta t + f(t_3) \Delta t.$$

Here t_i is the position of the ith rectangle along the t axis.

To get the exact answer, we take the limit in which Δt shrinks to zero and the number of rectangles increases to infinity. That defines the *definite integral* of $f(t)$ between the limits $t = a$ and $t = b$. We write this as

$$A = \int_a^b f(t)\, d t = \lim_{\Delta t \to 0} \sum_i f(t_i) \Delta t.$$

The integral sign, called *summa*, $\int$, replaces the summation sign, and—as in differential calculus—Δt is replaced by $d t$. The function $f(t)$ is called the *integrand*.

Let's make a notational change and call one of the limits of integration T. In particular, replace b by T and consider the integral

$$\int_a^T f(t)\, d t$$

where we are going to think of T as a variable instead of as a definite value of t. In this case, this integral defines a function of T, which can take on any value of t. The integral is a function of

T because it has a definite value for each value of T.

$$F(T) = \int_a^T f(t)\, d\,t.$$

Thus a given function $f(t)$ defines a second function $F(T)$. We could also let a vary, but we won't. The function $F(T)$ is called the *indefinite integral* of $f(t)$. It is indefinite because instead of integrating from a fixed value to a fixed value, we integrate to a variable. We usually write such an integral without limits of integration,

$$F(t) = \int f(t)\, d\,t. \tag{1}$$

The fundamental theorem of calculus is one of the simplest and most beautiful results in mathematics. It asserts a deep connection between integrals and derivatives. What it says is that if $F(T) = \int f(t)\, d\,t$, then

$$f(t) = \frac{d\,F(t)}{d\,t}.$$

To see this, consider a small incremental change in T from T to $T + \Delta\,t$. Then we have a new integral,

$$F(T + \Delta\,t) = \int_a^{T+\Delta\,t} f(t)\, d\,t.$$

In other words, we have added one more rectangle of width $\Delta\,t$ at $t = T$ to the area shaded in Figure 3. In fact, the difference $F(T + \Delta\,t) - F(T)$ is just the area of that extra rectangle, which happens to be $f(T)\,\Delta\,t$, so

$$F(T + \Delta\,t) - F(T) = f(T)\,\Delta\,t.$$

Dividing by Δt,

$$\frac{F(T + \Delta t) - F(T)}{\Delta t} = f(T)$$

we obtain the fundamental theorem connecting F and f, when we take the limit where $\Delta t \to 0$:

$$\frac{dF}{dT} = \lim_{\Delta t \to 0} \frac{F(T + \Delta t) - F(T)}{\Delta t} = f(T).$$

We can simplify the notation by ignoring the difference between t and T,

$$\frac{dF}{dt} = f(t).$$

In other words, the processes of integration and differentiation are reciprocal: The derivative of the integral is the original integrand.

Can we completely determine $F(t)$ knowing that its derivative is $f(t)$? Almost, but not quite. The problem is that adding a constant to $F(t)$ does not change its derivative. Given $f(t)$, its indefinite integral is ambiguous, but only up to adding a constant.

To see how the fundamental theorem is used, let's work out some indefinite integrals. Let's find the integral of a power $f(t) = t^n$. Consider,

$$F(t) = \int f(t)\, dt.$$

It follows that

$$f(t) = \frac{d\,F(t)}{d\,t}$$

or

$$t^n = \frac{d\,F(t)}{d\,t}.$$

All we need to do is find a function F whose derivative is t^n, and that is easy.

In the last chapter we found that for any m,

$$\frac{d\,(t^m)}{d\,t} = m\,t^{m-1}.$$

If we substitute $m = n + 1$, this becomes

$$\frac{d\left(t^{n+1}\right)}{d\,t} = (n+1)\,t^n$$

or, dividing by $n + 1$,

$$\frac{d\left(t^{n+1} \,/\, n+1\right)}{d\,t} = t^n.$$

Thus we find that t^n is the derivative of $\frac{t^{n+1}}{n+1}$. Substituting the relevant values, we get

$$F(t) = \int t^n\,d\,t = \frac{t^{n+1}}{n+1}.$$

The only thing missing is the ambiguous constant that we can add to F. We should write

$$\int t^n \, dt = \frac{t^{n+1}}{n+1} + c$$

where c is a constant that has to be determined by other means.

The ambiguous constant is closely related to the ambiguity in choosing the other endpoint of integration that we earlier called a. To see how a determines the ambiguous constant c, let's consider the integral

$$\int_a^T f(t) \, dt.$$

in the limit where the two limits of integration come together—that is, $T = a$. In this case, the integral has to be zero. You can use that fact to determine c.

In general, the fundamental theorem of calculus is written

$$\int_a^b f(t) \, dt = F(t) \Big|_a^b = F(b) - F(a). \tag{2}$$

Another way to express the fundamental theorem is by a single equation:

$$\int \frac{d f}{d t} \, dt = f(t) + c. \tag{3}$$

In other words, integrating a derivative gives back the original function (up to the usual ambiguous constant). Integration and differentiation undo each other.

Here are some integration formulas:

$$\int c \, dt = c t$$

$$\int c f(t)\, dt = c \int f(t)\, dt$$

$$\int t\, dt = \frac{t^2}{2} + c$$

$$\int t^2\, dt = \frac{t^3}{3} + c$$

$$\int t^n\, dt = \frac{t^{n+1}}{n+1} + c$$

$$\int \sin t\, dt = -\cos t + c$$

$$\int \cos t\, dt = \sin t + c$$

$$\int e^t\, dt = e^t$$

$$\int \frac{dt}{t} = \ln t + c$$

$$\int [f(t) \pm g(t)]\, dt = \int f(t)\, dt \pm \int g(t)\, dt.$$

Exercise 1: Determine the indefinite integral of each of the following expressions by reversing the process of differentiation and adding a constant.

$f(t) = t^4$

$f(t) = \cos t$

$f(t) = t^2 - 2$

Exercise 2: Use the fundamental theorem of calculus to evaluate each integral from Exercise 1 with limits of integration being $t = 0$ to $t = T$.

Exercise 3: Treat the expressions from Exercise 1 as expressions for the acceleration of a particle. Integrate them once, with respect to time, and determine the velocities, and a second time to determine the trajectories. Because we will use t as one of the limits of integration we will adopt the dummy integration variable t'. Integrate them from $t' = 0$ to $t' = t$.

$$v(t) = \int_0^t t'^4 \, d\,t'$$
$$v(t) = \int_0^t \cos t' \, d\,t'$$
$$v(t) = \int_0^t \left(t'^2 - 2\right) d\,t'$$

Integration by Parts

There are some tricks to doing integrals. One trick is to look them up in a table of integrals. Another is to learn to use *Mathematica*. But if you're on your own and you don't recognize the integral, the oldest trick in the book is *integration by parts*. It's just the reverse of using the product rule for differentiation. Recall from Lecture 2 that to differentiate a function, which itself is a product of two functions, you use the following rule:

$$\frac{d[f(x)\,g(x)]}{d\,x} = f(x)\,\frac{d\,g(x)}{d\,x} + g(x)\,\frac{d\,f(x)}{d\,x}.$$

Now let's integrate both sides of this equation between limits a

and b.

$$\int_a^b \frac{d[f(x)\,g(x)]}{d\,x} = \int_a^b f(x)\,\frac{d\,g(x)}{d\,x} + \int_a^b g(x)\,\frac{d\,f(x)}{d\,x}$$

The left side of the equation is easy. The integral of a derivative (the derivative of $f\,g$) is just the function itself. The left side is

$$f(b)\,g\,(b) - f(a)\,g(a)$$

which we often write in the form

$$f(x)\,g\,(x)|_a^b.$$

Now let's subtract one of the two integrals on the right side and shift it to the left side.

$$f(x)\,g\,(x)|_a^b - \int_a^b f(x)\,\frac{d\,g(x)}{d\,x} = \int_a^b g(x)\,\frac{d\,f(x)}{d\,x}. \tag{4}$$

Suppose we have some integral that we don't recognize, but we notice that the integrand happens to be a product of a function $f(x)$ and the derivative of another function $g(x)$. In other words, after some inspection, we see that the integral has the form of the right side of Eq. (4), but we don't know how to do it. Sometimes we are lucky and recognize the integral on the left side of the equation.

Let's do an example. Suppose the integral that we want to do is

$$\int_0^{\frac{\pi}{2}} x \cos x\,d\,x.$$

That's not in our list of integrals. But notice that

$$\cos x = \frac{d \sin x}{d x}$$

so the integral is

$$\int_0^{\frac{\pi}{2}} x \, \frac{d \sin x}{d x} \, d x.$$

Equation (4) tells us that this integral is equal to

$$x \sin x \Big|_0^{\frac{\pi}{2}} - \int_0^{\frac{\pi}{2}} \frac{d x}{d x} \sin x \, d x$$

or just

$$\frac{\pi}{2} \sin \frac{\pi}{2} - \int_0^{\frac{\pi}{2}} \sin x \, d x.$$

Now it's easy. The integral $\int \sin x \, d x$ is on our list: it's just $\cos x$. I'll leave the rest to you.

Exercise 4: Finish evaluating $\int_0^{\frac{\pi}{2}} x \cos x \, d x$.

You might wonder how often this trick works. The answer is quite often, but certainly not always. Good luck.

Lecture 3: Dynamics

Lenny: "What makes things move, George?

 George: "Forces do, Lenny."

 Lenny: "What makes things stop moving, George?"

 George: "Forces do, Lenny."

Aristotle's Law of Motion

Aristotle lived in a world dominated by friction. To make anything move—a heavy cart with wooden wheels, for example—you had to push it, you had to apply a *force* to it. The harder you pushed it, the faster it moved; but if you stopped pushing, the cart very quickly came to rest. Aristotle came to some wrong conclusions because he didn't understand that friction is a force. But still, it's worth exploring his ideas in modern language. If he had known calculus, Aristotle might have proposed the following law of motion:

The velocity of any object is proportional to the total applied force.

Had he known how to write vector equations, his law would have looked like this:

$$\vec{F} = m\vec{v}.$$

$\vec{F}$ is of course the applied force, and the response (according to Aristotle) would be the velocity vector, $\vec{v}$. The factor m relating the two is some characteristic quantity describing the resistance of the body to being moved; for a given force, the bigger the m of the object, the smaller its velocity. With a little reflection, the old philosopher might have identified m with the mass of the object. It would have been obvious that heavier things are harder to move than lighter things, so somehow the mass of the object has to be in the equation.

One suspects that Aristotle never went ice skating, or he would have known that it is just as hard to stop a body as to get it moving. Aristotle's law is just plain wrong, but it is nevertheless worth studying as an example of how equations of motion can determine the future of a system. From now on, let's call the body a particle.

Consider one-dimensional motion of a particle along the x axis under the influence of a given force. What I mean by a given force is simply that we know what the force is at any time. We can call it $F(t)$ (note that vector notation would be a bit redundant in one dimension). Using the fact that the velocity is the time derivative of position, x, we find that Aristotle's equation takes the form

$$\frac{dx(t)}{dt} = \frac{F(t)}{m}.$$

Before solving the equation, let's see how it compares to the deterministic laws of Chapter 1. One obvious difference is that Aristotle's equation is not stroboscopic—that is neither t nor x is discrete. They do not change in sudden stroboscopic steps; they change continuously. Nevertheless, we can see the similarity if we assume that time is broken up into intervals of size Δt and replace the derivative by $\frac{\Delta x}{\Delta t}$. Doing so gives

$$x(t + \Delta t) = x(t) + \Delta t \frac{F(t)}{m}.$$

In other words, wherever the particle happens to be at time t, at the next instant its position will have shifted by a definite amount. For example, if the force is constant and positive, then in each incremental step the particle moves forward by an amount $\Delta t \frac{F(t)}{m}$. This law is obviously deterministic. Knowing that the particle was at a point $x(0)$ at time $t = 0$ (or x_0), one can easily predict where it will be in the future. So by the criteria of Chapter 1, Aristotle did not commit any crime.

Let's go back to the exact equation of motion:

$$\frac{dx(t)}{dt} = \frac{F(t)}{m}.$$

Equations for unknown functions that involve derivatives are called *differential equations*. This one is a *first-order* differential equation because it contains only first derivatives. Equations like this are easy to solve. The trick is to integrate both sides of the equation:

$$\int \frac{dx(t)}{dt} dt = \int \frac{F(t)}{m} dt.$$

The left side of the equation is the integral of a derivative. That's where the fundamental theorem of calculus comes in handy. The left side is just $x(t) + c$.

The right side, on the other hand, is the integral of some specified function and, apart from a constant, is also determined. For example, if F is constant, then the right side is

$$\int \frac{F}{m} dt = \frac{F}{m} t + c.$$

Note that we included an additive constant. Putting an arbitrary constant on both sides of the equation is redundant. In this case, the equation of motion is satisfied by

$$x(t) = \frac{F}{m} t + c.$$

How do you fix the constant c? The answer is by the initial condition. For example, if we knew that the particle started at $x = 1$, at time $t = 3$ we would plug these values in, obtaining

$$1 = \frac{F}{m} 3 + c,$$

and solve for c:

$$c = 1 - 3 \frac{F}{m}.$$

Exercise 1: Given a force that varies with time according to $F = 2t^2$, and with the initial condition at time zero, $x(0) = \pi$, use Aristotle's law to find $x(t)$ at all times.

Aristotle's equations of motion are deterministic, but are they reversible? In Lecture 1, I explained that *reversible* means that if all the arrows were reversed, the resulting new law of motion would also be deterministic. The analogous procedure to reversing the arrows when time is continuous is very simple. Everywhere you see time in the equations, replace it with minus time. That will have the effect of interchanging the future and the past. Changing t to $-t$ also includes changing the sign of small differences in time. In other words, every Δt must be replaced with $-\Delta t$. In fact, you can do it right at the level of the differentials dt. *Reversing the arrows* means changing the differential dt to $-dt$. Let's go back to Aristotle's equation

$$F(t) = m\frac{dx}{dt}$$

and change the sign of time. The result is

$$F(-t) = -m\frac{dx}{dt}.$$

The left-hand side of the equation is the force, but the force evaluated at time $-t$, not at time t. However, if $F(t)$ is a known function, then so is $F(-t)$. In the reversed problem, the force is also a known function of reversed time.

On the right-hand side of the equation we've replaced dt with $-dt$, thereby changing the sign of the whole expression. In fact, one can shift the minus sign to the left-hand side of the equation:

$$-F(-t) = m\frac{dx}{dt}.$$

The implication is simple: The reversed equation of motion is exactly like the original, but with a different rule for the force as a function of time. The conclusion is clear: If Aristotle's equations of motion are deterministic into the future, they are also deterministic into the past. The problem with Aristotle's equations is not that they are inconsistent; they are just the wrong equations.

It is interesting that Aristotle's equations do have an application—not as fundamental laws, but as approximations. Frictional forces do exist, and in many cases they are so important that Aristotle's intuition—things stop if you stop pushing—is almost correct. Frictional forces are not fundamental. They are a consequence of a body interacting with a huge number of other tiny bodies—atoms and molecules—that are too small and too numerous to keep track of. So we average over all the hidden degrees of freedom. The result is frictional forces. When frictional forces are very strong such as in a stone moving through mud—then Aristotle's equation is a very good approximation, but with a qualification. It's not the mass that determines the proportionality between force and velocity. It's the so-called viscous drag coefficient. But that may be more than you want to know.

Mass, Acceleration, and Force

Aristotle's mistake was to think that a net "applied" force is needed to keep an object moving. The right idea is that one force—the applied force—is needed to overcome another force—the force of friction. An isolated object moving in free space, with no forces acting on it, requires nothing to keep it moving. In fact, it needs a force to stop it. This is the *law of inertia*. What forces do is change the state of motion of a body. If the body is initially at rest, it takes a force to start it moving. If it's moving, it takes a force to stop it. If it is moving in a particular direction, it takes a force to change the direction of motion. All of these examples involve a change in the velocity of an object, and therefore an acceleration.

From experience we know that some objects have more inertia than others; it requires a larger force to change their velocities. Obvious examples of objects possessing large and small inertia are locomotives and Ping-Pong balls, respectively. The quantitative measure of an object's inertia is its *mass*.

Newton's law of motion involves three quantities: acceleration, mass, and force. Acceleration we studied in Lecture 2. By monitoring the position of an object as it moves, a clever observer—with a bit of mathematics—can determine its acceleration. Mass is a new concept that is actually defined in terms of force and acceleration. But so far we haven't defined force. It sounds like we are in a logical circle in which force is defined by the ability to change the motion of a given mass, and mass is defined by the resistance to that change. To break that circle, let's take a closer look at how force is defined and measured in practice.

There are very sophisticated devices that can measure force to great accuracy, but it will suit our purposes best to imagine a very old-fashioned device, namely, a spring balance. It consists of a spring and a ruler to measure how much the spring is stretched from its natural equilibrium length (see Figure 1).

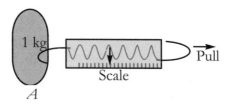

Figure 1: A spring balance.

The spring has two hooks, one to attach to the massive body whose mass is being measured, and one to pull on. In fact, while you are at it, make several such identical devices.

Let's define a unit of force by pulling on one hook, while holding the other hook fixed to some object *A*, until the pointer registers one "tick" on the ruler. Thus we are applying a unit of force to *A*.

To define two units of force, we could pull just hard enough to stretch the spring to two ticks. But this assumes that the spring behaves the same way between one tick and two ticks of stretching. This will lead us back to a vicious circle of reasoning that we don't want to get into. Instead, we define two units of force by attaching two spring balances to *A* and pulling both of them with a single unit of force (see Figure 2).

In other words, we pull both hooks so that each pointer records a single tick. Three units of force would be defined by using three springs, and so on.

When we do this experiment in free space, we discover the interesting fact that object *A* accelerates along the direction in which we pull the hook. More exactly, the acceleration is proportional to the force—twice as big for two units of force, three times as big for three units, and so on.

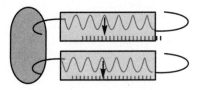

Figure 2: Twice the force.

Let us do something to change the inertia of *A*. In particular, we will double the inertia by hooking together two identical versions of object *A* (see Figure 3).

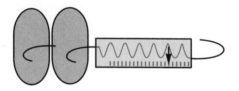

Figure 3: Twice the mass.

What we find is that when we apply a single unit of force (by pulling the whole thing with a single spring stretched to one tick) the acceleration is only half what it was originally. The inertia (mass) is now twice as big as before.

The experiment can obviously be generalized; hook up three masses, and the acceleration is only a third as big, and so on.

We can do many more experiments in which we hook any number of springs to any number of A's. The observations are summarized by a single formula, Newton's second law of motion, which tells us that force equals mass times acceleration,

$$\vec{F} = m\vec{a}. \tag{1}$$

This equation can also be written in the form

$$\vec{F} = m\frac{d\vec{v}}{dt}. \tag{2}$$

In other words, **force equals mass times the rate of change of velocity**: no force—no change in velocity.

Note that these equations are vector equations. Both force and acceleration are vectors because they not only have magnitude but also direction.

An Interlude on Units

A mathematician might be content to say that the length of a line segment is 3. But a physicist or engineer—or even an ordinary person—would want to know, "Three what?" Three inches, three centimeters, or three light years?

Similarly, it conveys no information to say that the mass of an object is 7 or 12. To give the numbers meaning, we must indicate what units we are using. Let's begin with length.

Somewhere in Paris rests the defining platinum meter stick. It is kept in a sealed container at a fixed temperature and away from other conditions that might affect its length.[1] From here on, we will adopt that meter stick as our unit of length.

Thus we write

$$[x] = [\text{length}] = \text{meters}.$$

Despite its appearance, this is not an equation in the usual sense. The way to read it is *x has units of length and is measured in meters.*

Similarly, *t* has units of time and is measured in seconds. The definition of a second could be given by the amount of time it takes a certain pendulum to make a single swing:

$$[t] = [\text{time}] = \text{seconds}.$$

The units meters and seconds are abbreviated as m and s, respectively.

---

[1] There is a more modern definition of the meter in terms of the wavelength of light emitted by atoms jumping from one quantum level to another. For our purposes the Paris meter stick will do just fine.

Once we have units for length and time, we can construct units for velocity and acceleration. To compute the velocity of an object, we divide a distance by a time. The result has units of *length per time*, or—in our units—meters per second.

$$[v] = \left[\frac{\text{length}}{\text{time}}\right] = \frac{\text{m}}{\text{s}}.$$

Similarly, acceleration is the rate of change of velocity, and its units are velocity per unit time, or length per unit time per unit time:

$$[a] = \left[\frac{\text{length}}{\text{time}}\right]\left[\frac{1}{\text{time}}\right] = \left[\frac{\text{length}}{\text{time}^2}\right] = \frac{\text{m}}{\text{s}^2}.$$

The unit of mass that we will use is the kilogram; it is defined as the mass of a certain lump of platinum, that is also kept somewhere in France. Thus

$$[m] = [\text{mass}] = \text{kilogram} = \text{kg}.$$

Now let's consider the unit of force. One might define it in terms of some particular spring made of a specific metal, stretched a distance of 0.01 meter, or something like that. But in fact, we have no need for a new unit of force. We already have one—namely the force that it takes to accelerate one kilogram by one meter per second per second. Even better is to use Newton's law $F = ma$. Evidently, force has units of mass times acceleration,

$$[F] = [\text{force}]$$

$$= [ma]$$

$$= \left[\frac{\text{mass} \times \text{length}}{\text{time}^2} \right]$$

$$= \frac{\text{kg m}}{\text{s}^2}.$$

There is a name for this unit of force. One kilogram meter per second squared is called a Newton, abbreviated N. Newton, himself, being English, probably favored the British unit, namely the pound. There are about 4.4 N to a pound.

Some Simple Examples of Solving Newton's Equations

The simplest of all examples is a particle with no forces acting on it. The equation of motion is Eq. (2), but with the force set to zero:

$$m \frac{d \vec{v}}{dt} = 0,$$

or, using the dot notation for time derivative,

$$m \dot{\vec{v}} = 0.$$

We can drop the factor of mass and write the equation in component form as

$$\dot{v}_x = 0$$
$$\dot{v}_y = 0$$
$$\dot{v}_z = 0$$

The solution is simple: The components of velocity are constant and can just be set equal to their initial values,

$$v_x(t) = v_x(0). \tag{3}$$

The same goes for the other two components of velocity. This, incidently, is often referred to as *Newton's first law of motion*:

Every object in a state of uniform motion tends to remain in that state of motion unless an external force is applied to it.

Equations (1) and (2) are called *Newton's second law of motion*,

The relationship between an object's mass m, its acceleration a, and the applied force F is

$$F = ma.$$

But, as we have seen, the first law is simply a special case of the second law when the force is zero.

Recalling that velocity is the derivative of position, we can express Eq. (3) in the form

$$\dot{x} = v_x(0).$$

This is the simplest possible differential equation, whose solution (for all components) is

$$x(t) = x_0 + v_x(0)t$$
$$y(t) = y_0 + v_y(0)t$$
$$z(t) = z_0 + v_z(0)t$$

or, in vector notation,

$$\vec{r}(t) = \vec{r}_0 + \vec{v}_0 t.$$

A more complicated motion results from the application of a constant force. Let's first carry it out for just the z direction. Dividing by m, the equation of motion is

$$\dot{v}_z = \frac{F_z}{m}.$$

> **Exercise 2: Integrate this equation.** *Hint: Use definite integrals.*

From this result we deduce

$$v_z(t) = v_z(0) + \frac{F_z}{m}t,$$

or

$$\dot{z}(t) = v_z(0) + \frac{F_z}{m}t.$$

This is probably the second simplest differential equation. It is easy to solve:

$$z(t) = z_0 + v_z(0)t + \frac{F_z}{2m}t^2. \tag{4}$$

> **Exercise 3: Show by differentiation that this satisfies the equation of motion.**

This simple case may be familiar. If z represents the height above the surface of the Earth, and $\dfrac{F_z}{m}$ is replaced with the acceleration due to gravity, $\dfrac{F_z}{m} = -g$, then Eq. (4) is the equation describing the motion of an object falling from height z_0 with an initial velocity $v_z(0)$:

$$z(t) \;=\; z_0 + v_z(0)t - \frac{1}{2}gt^2. \tag{5}$$

Let's consider the case of the simple harmonic oscillator. This system is best thought of as a particle that moves along the x axis, subject to a force that pulls it toward the origin. The force law is

$$F_x = -kx.$$

The negative sign indicates that at whatever the value of x, the force pulls it back toward $x = 0$. Thus, when x is positive, the force is negative, and vice versa. The equation of motion can be written in the form

$$\ddot{x} = -\frac{k}{m}x,$$

or, by defining $\frac{k}{m} = \omega^2$,

$$\ddot{x} = -\omega^2 x. \tag{6}$$

Exercise 4: Show by differentiation that the general solution to Eq. (6) is given in terms of two constants A and B by

$$x(t) = A \cos \omega t + B \sin \omega t.$$

Determine the initial position and velocity at time $t = 0$ in terms of A and B.

The harmonic oscillator is an enormously important system that occurs in contexts ranging from the motion of a pendulum to the oscillations of the electric and magnetic fields in a light wave. It is profitable to study it thoroughly.

Interlude 3: Partial Differentiation

"Look out there, Lenny. Ain't those hills and valleys pretty?"

"Yeah George. Can we get a place over there when we get some money? Can we?"

George squinted. "Exactly where are you looking Lenny?"

Lenny pointed. "Right over there George. That local minimum."

Partial Derivatives

The calculus of multivariable functions is a straightforward generalization of single-variable calculus. Instead of a function of a single variable t, consider a function of several variables. To illustrate, let's call the variables x, y, z, although these don't have to stand for the coordinates of ordinary space. Moreover, there can be more or fewer then three. Let us also consider a function of these variables, $V(x, y, z)$. For every value of x, y, z, there is a unique value of $V(x, y, z)$ that we assume varies smoothly as we vary the coordinates.

Multivariable differential calculus revolves around the concept of *partial derivatives*. Suppose we are examining the neighborhood of a point x, y, z, and we want to know the rate at which V varies as we change x while keeping y and z fixed. We can just imagine that y and z are fixed parameters, so the only variable is x. The derivative of V is then defined by

$$\frac{dV}{dx} = \lim_{\Delta x \to 0} \frac{\Delta V}{\Delta x} \tag{1}$$

where ΔV is defined by

$$\Delta V = V([x + \Delta x], y, z) - V(x, y, z). \qquad (2)$$

Note that in the defininition of ΔV, only x has been shifted; y and z are kept fixed.

The derivative defined by Eq. (1) and Eq. (2) is called the *partial derivative* of V with respect to x and is denoted

$$\frac{\partial V}{\partial x}$$

or, when we want to emphasize that y and z are kept fixed,

$$\left(\frac{\partial V}{\partial x}\right)_{y,z}.$$

By the same method we can construct the partial derivative with respect to either of the other variables:

$$\frac{\partial V}{\partial y} = \lim_{\Delta y \to 0} \frac{\Delta V}{\Delta y}.$$

A shorthand notation for the partial derivatives of V with respect to y is

$$\frac{\partial V}{\partial y} = \partial_y V.$$

Multiple derivatives are also possible. If we think of $\dfrac{\partial V}{\partial x}$ as itself being a function of x, y, z, then it can be differentiated. Thus we can define the second-order partial derivative with respect to x:

$$\frac{\partial^2 V}{\partial x^2} = \partial_x \left(\frac{\partial V}{\partial x} \right) = \partial_{x,x} V.$$

Mixed partial derivatives also make sense. For example, one can differentiate $\partial_y V$ with respect to x:

$$\frac{\partial^2 V}{\partial x \partial y} = \partial_x \left(\frac{\partial V}{\partial y} \right) = \partial_{x,y} V.$$

It's an interesting and important fact that the mixed derivatives do not depend on the order in which the derivatives are carried out. In other words,

$$\frac{\partial^2 V}{\partial x \partial y} = \frac{\partial^2 V}{\partial y \partial x}.$$

Exercise 1: Compute all first and second partial derivatives—including mixed derivatives—of the following functions.

$x^2 + y^2 = \sin(x\,y)$

$\frac{x}{y}\, e^{x^2 + y^2}$

$e^x \cos y$

Stationary Points and Minimizing Functions

Let's look at a function of y that we will call F (see Figure 1).

$F(y)$

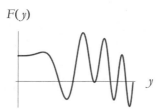

Figure 1: Plot of the function $F(y)$.

Notice that there are places on the curve where a shift in y in either direction produces only an upward shift in F. These points are called *local minima*. In Figure 2 we have added dots to indicate the local minima.

$F(y)$

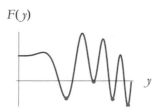

Figure 2: Local minima.

For each local minimum, when you go in either direction along y, you begin to rise above the dot in $F(y)$. Each dot is at the bottom of a little depression. The *global minimum* is the lowest possible place on the curve.

One condition for a local minimum is that the derivative of the function with respect to the independent variable at that point is zero. This is a necessary condition, but not a sufficient condition. This condition defines any *stationary point*,

$$\frac{d}{dy} F(y) = 0.$$

The second condition tests to see what the character of the stationary point is by examining its second derivative. If the second derivative is larger than 0, then all points nearby will be above the stationary point, and we have a *local minimum*:

$$\frac{d^2}{d^2 y} F(y) > 0.$$

If the second derivative is less than 0, then all points nearby will be below the stationary point, and we have a *local maximum*:

$$\frac{d^2}{d^2 y} F(y) < 0.$$

See Figure 3 for examples of local maxima.

$F(y)$

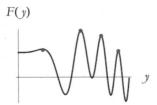

Figure 3: Local maxima.

If the second derivative is equal to 0, then the derivative changes from positive to negative at the stationary point, which we call a *point of inflection*:

$$\frac{d^2}{d^2 y} F(y) = 0.$$

See Figure 4 for an example of a point of inflection.

$$F(y)$$

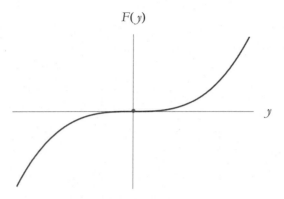

y

Figure 4: Point of inflection.

These are collectively the results of a *second-derivative test*.

Stationary Points in Higher Dimensions

Local maxima, local minima, and other stationary points can happen for functions of more than one variable. Imagine a hilly terrain. The altitude is a function that depends on the two coordinates—let's say latitude and longitude. Call it $A(x, y)$. The tops of hills and the bottoms of valleys are local maxima and minima of $A(x, y)$. But they are not the only places where the terrain is locally horizontal. Saddle points occur between two hills. You can see some examples in Figure 5.

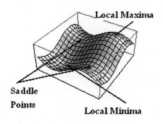

Figure 5: A function of several variables.

The very tops of hills are places where no matter which way you move, you soon go down. Valley bottoms are the opposite; all directions lead up. But both are places where the ground is level.

There are other places where the ground is level. Between two hills you can find places called saddles. *Saddle points* are level, but along one axis the altitude quickly increases in either direction. Along another perpendicular direction the altitude decreases. All of these are called stationary points.

Let's take a slice along the x axis through our space so that the slice passes through a local minimum of A, see Figure 6.

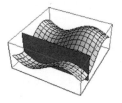

Figure 6: A slice along the x axis.

It's apparent that at the minimum, the derivative of A with respect to x vanishes, we write this:

$$\frac{\partial A}{\partial x} = 0.$$

On the other hand, the slice could have been oriented along the y axis, and we would then conclude that

$$\frac{\partial A}{\partial y} = 0.$$

To have a minimum, or for that matter to have any stationary point, both derivatives must vanish. If there were more directions of space in which A could vary, then the condition for a stationary point is given by:

$$\frac{\partial A}{\partial x_i} = 0. \tag{3}$$

for all x_i.

There is a shorthand for summarizing these equations. Recall that the change in a function when the point x is varied a little bit is given by

$$\delta A = \sum_i \frac{\partial A}{\partial x_i} \delta x_i.$$

The set of Equations (3) are equivalent to the condition that

$$\delta A = 0 \tag{4}$$

for any small variation of x.

Suppose we found such a point. How do we tell whether it is a maximum, a minimum, or a saddle. The answer is a generalization of the criterion for a single variable. We look at the second derivatives. But there are several second derivatives. For the case of two dimensions, we have

$$\frac{\partial^2 A}{\partial x^2},$$

$$\frac{\partial^2 A}{\partial y^2},$$

$$\frac{\partial^2 A}{\partial x \partial y},$$

and

$$\frac{\partial^2 A}{\partial y \partial x},$$

the last two being the same.

These partial derivatives are often arranged into a special matrix called the *Hessian matrix.*

$$H = \begin{pmatrix} \dfrac{\partial^2 A}{\partial x^2} & \dfrac{\partial^2 A}{\partial x \partial y} \\ \dfrac{\partial^2 A}{\partial y \partial x} & \dfrac{\partial^2 A}{\partial y^2} \end{pmatrix}.$$

Important quantities, called the determinant and the trace, can be made out of such a matrix. The determinant is given by

$$\text{Det } H = \frac{\partial^2 A}{\partial x^2} \frac{\partial^2 A}{\partial y^2} - \frac{\partial^2 A}{\partial y \partial x} \frac{\partial^2 A}{\partial x \partial y}$$

and the trace is given by

$$\text{Tr } H = \frac{\partial^2 A}{\partial x^2} + \frac{\partial^2 A}{\partial y^2}.$$

Matrices, determinants, and traces may not mean much to you beyond these definitions, but they will if you follow these lectures to the next subject—quantum mechanics. For now, all you need is the definitions and the following rules.

If the determinant and the trace of the Hessian is positive then the point is a local minimum.

If the determinant is positive and the trace negative the point is a local maximum.

If the determinant is negative, then irrespective of the trace, the point is a saddle point.

However: One caveat, these rules specifically apply to functions of two variables. Beyond that, the rules are more complicated.

None of this is obvious for now, but it still enables you to test various functions and find their different stationary points. Let's take an example. Consider

$$F(x, y) = \sin x + \sin y.$$

Differentiating, we get

$$\frac{\partial F}{\partial x} = \cos x$$

$$\frac{\partial F}{\partial y} = \cos y.$$

Take the point $x = \frac{\pi}{2}$, $y = \frac{\pi}{2}$. Since $\cos \frac{\pi}{2} = 0$, both derivatives are zero and the point is a stationary point.

Now, to find the type of stationary point, compute the second derivatives. The second derivatives are

$$\frac{\partial^2 F}{\partial x^2} = -\sin x$$

$$\frac{\partial^2 F}{\partial y^2} = -\sin y$$

$$\frac{\partial^2 F}{\partial x \partial y} = 0$$

$$\frac{\partial^2 F}{\partial y \partial x} = 0.$$

Since $\sin \frac{\pi}{2} = 1$ we see that both the determinant and the trace of the Hessian are positive. The point is therefore a minimum.

Exercise 2: Consider the points $\left(x = \frac{\pi}{2}, y = -\frac{\pi}{2}\right)$, $\left(x = -\frac{\pi}{2}, y = \frac{\pi}{2}\right)$, $\left(x = -\frac{\pi}{2}, y = -\frac{\pi}{2}\right)$. **Are these points stationary points of the following functions? Is so, of what type?.**

$F(x, y) = \sin x + \sin y$

$F(x, y) = \cos x + \cos y$

Lecture 4: Systems of More Than One Particle

It's a lazy, warm evening. Lenny and George are lying in the grass looking up at the sky.

"Tell me about the stars George. Are they particles?"

"Kind of, Lenny."

"How come they don't move?"

"They do, Lenny. It's just that they're very far away."

"There's an awful lot of them, George. Do you think that guy Laplace could really figure them all out?"

Systems of Particles

If—as Laplace believed—natural systems are composed of particles, then the laws of nature must be the dynamical laws of motion that determine the motion of those systems of particles. Again, Laplace: "An intellect which at a certain moment would know all forces . . . and all positions . . . ." What is it that determines the force on a given particle? It is the positions of all the other particles.

There are many types of forces—such as friction, the drag force exerted by the wind, and the force exerted by the floor that keeps you from falling to the basement—that are not fundamental. They originate from the microscopic interactions between atoms and molecules.

The fundamental forces are those that act between particles, like gravity and electric forces. These depend on a number of things: Gravitational forces between particles are proportional to the product of their masses, and electric forces are proportional to the product of their electric charges. Charges and masses are considered to be intrinsic properties of a particle, and specifying them is part of specifying the system itself.

Apart from the intrinsic properties, the forces depend on the location of the particles. For example, the distance between objects determines the electric and gravitational force that one particle exerts on another. Suppose that the locations of all the particles are described by their coordinates: x_1, y_1, z_1 for the first particle, x_2, y_2, z_2 for the second particle, x_3, y_3, z_3 for the third particle, and so on up to the last, or the Nth, particle. Then the force on any one particle is a function of its location as well as the location of all the others. We can write this in the form

$$\vec{F}_i = \vec{F}_i\left(\left\{\vec{r}\right\}\right).$$

What this equation means is that the force on the ith particle is a function of the positions of all the particles. The symbol $\left\{\vec{r}\right\}$ stands for the collective location of every particle in the system. Another way of saying this is that the symbol represents the set of all position vectors.

Once we know the force on any particle—for example, particle number 1—we can write Newton's equation of motion for that particle:

$$\vec{F}_1\left(\left\{\vec{r}\right\}\right) = m_1\,\vec{a}_1,$$

where m_1 and $\vec{a}_1$ are the mass and acceleration of particle 1. When we express the acceleration as the second derivative of the position, the equation becomes

$$\vec{F}_1\left(\left\{\vec{r}\right\}\right) = m_1 \frac{d^2 \vec{r}_1}{d t^2}.$$

In fact, we can write such an equation for each particle:

$$\vec{F}_1\left(\left\{\vec{r}\right\}\right) = m_1 \frac{d^2 \vec{r}_1}{d t^2}$$

$$\vec{F}_2\left(\left\{\vec{r}\right\}\right) = m_2 \frac{d^2 \vec{r}_2}{d t^2}$$

$$\vec{F}_3\left(\left\{\vec{r}\right\}\right) = m_3 \frac{d^2 \vec{r}_3}{d t^2}$$

$$\vdots$$

$$\vec{F}_N\left(\left\{\vec{r}\right\}\right) = m_N \frac{d^2 \vec{r}_N}{d t^2}$$

or, in condensed form,

$$\vec{F}_i\left(\left\{\vec{r}\right\}\right) = m_i \frac{d^2 \vec{r}_i}{d t^2}.$$

We can also write these equations in component form:

$$(F_x)_i\left(\{x\}\right) = m_i \frac{d^2 x_i}{d t^2}$$

$$\left(F_y\right)_i\left(\{y\}\right) = m_i \frac{d^2 y_i}{d t^2} \tag{1}$$

$$\left(F_z\right)_i\left(\{z\}\right) = m_i \frac{d^2 z_i}{d t^2}.$$

In this set of equations, $(F_x)_i$, $\left(F_y\right)_i$, and $\left(F_z\right)_i$ mean the x, y, and z components of the force on the ith particle, and the symbols

{x}, {y}, and {z} represent the sets of all the x coordinates, all the y coordinates, and all the z coordinates of all the particles.

This last set of equations makes it clear that there is an equation for each coordinate of every particle, which would tell Laplace's vast intellect how every particle moves if the initial conditions were known. How many equations are there in all? The answer is three for each particle, so if there are N particles the grand total is $3N$ equations.

The Space of States of a System of Particles

The formal meaning of the state of a system is, "Everything you need to know (with perfect accuracy) to predict its future, given the dynamical law." Recall from Lecture 1, that the space of states, or state-space, is the collection of all possible states of the system. In the examples of Lecture 1, the state-space was typically a discrete collection of possibilities: H or T for the coin, 1 through 6 for the die, and so forth. In Aristotelian mechanics, assuming that the forces on an object are known, the state is specified by simply knowing the location of the object. In fact, from Aristotle's law, the force determines the velocity, and the velocity tells you where the particle will be at the next instant.

But Newton's law is different from Aristotle's: It tells you the acceleration, not the velocity. This means that to get started, you need to know not only where the particles are but also their velocities. Knowing the velocity tells you where the particle will be at the next instant, and knowing the acceleration tells you what the velocity will be.

All of this means that the state of a system of particles consists of more than just their current locations; it also includes their current velocities. For example, if the system is a single particle, its state consists of six pieces of data: the three components of its position and the three components of its velocity. We may express this by saying that the state is a point in a six-dimensional space of states labeled by axes x, y, z, v_x, v_y, v_z.

Now let's consider the motion of the particle. At each instant of time, the state is specified by the values of the six variables $x(t)$, $y(t)$, $z(t)$, $v_x(t)$, $v_y(t)$, $v_z(t)$. The history of the particle can be pictured as a trajectory through the six-dimensional state-space.

Next, consider the space of states of a system of N particles. To specify the state of the system, we need to specify the state of every particle. This obviously means that the space of states is $6N$-dimensional: three position components and three velocity components for each of the N particles. One may even say that the motion of the system is a trajectory through a $6N$-dimensional space.

But wait. If the state-space is $6N$-dimensional, why is it that $3N$ components in Equations (1) are enough to determine how the system evolves? Are we missing half the equations? Let's go back to a system of a single particle with specified forces and write Newton's equations, using the fact that acceleration is the rate of change of velocity:

$$m\,\frac{d\vec{v}}{dt} = \vec{F}.$$

Since there is no expression for the velocity here, let's add to this another equation expressing the fact that velocity is the rate of change of position:

$$\frac{d\vec{r}}{dt} = \vec{v}.$$

When we include this second equation, we have a total of six components that tell us how the six coordinates of the state-space change with time. The same idea, applied to each individual particle, gives us $6N$ equations governing the motion through the space of states:

$$m_i \frac{dv_i}{dt} = F_i$$

$$\frac{dr_i}{dt} = v_i. \tag{2}$$

Thus, in answer to the question posed above, we *were* missing half the equations.

Wherever you happen to be in the $6N$-dimensional space of states, Equations (2) tell you where you will be next. They also tell you where you were an instant ago. Thus, Equations (2) are suitable dynamical laws. We now have our $6N$ equations for the N particles.

Momentum and Phase Space

If you are struck by a moving object, the result depends not only on the velocity of the object but also on its mass. Obviously, a Ping-Pong ball at 30 miles per hour (about 13 meters per second) will have much less of a mechanical effect than a locomotive moving at the same speed. In fact, the effect is proportional to the *momentum* of the object, which for now we shall define as the product of the velocity and the mass. Since the velocity is a vector, so is the momentum, denoted by the letter p. Thus

$$p_i = m_i\, v_i$$

or

$$\vec{p} = m\, \vec{v}.$$

Since velocity and momentum are so closely linked, we can use momentum and position instead of velocity and position to label the points of the state-space. When the state-space is described this way, it has a special name—*phase space*. The phase space of a particle is a six-dimensional space with coordinates x_i and p_i (see Figure 1).

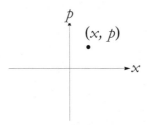

Figure 1: A point in phase space.

Why didn't we call this space configuration space? Why the new term *phase space*? The reason is that the term configuration space is used for something else, namely, the three-dimensional space of positions: Just the r_i's. It might have been called position space; then we could have said, "Position space plus momentum space equals phase space." In fact, we do say that, but we also use the term configuration space interchangeably with position space. Therefore the slogan is

Configuration space plus momentum space equals phase space.

You may wonder why we go to the trouble of replacing the intuitive concept of velocity with the more abstract concept of momentum in describing the state of a particle. The answer should become clear as we develop the basic framework of classical mechanics in later chapters. For now, let's just reexpress Equations (2) in terms of momentum instead of velocity. To do so, we first note that

$$m \, \frac{d\vec{v}}{dt}$$

is nothing but the time rate of change of momentum—that is, $\frac{d\vec{p}}{dt}$, or in the condensed dot notation,

$$m \, \frac{d\vec{v}}{dt} = \dot{\vec{p}}.$$

The full set of equations becomes

$$\dot{p}_i = F_i(\{r_i\})$$
$$\dot{r}_i = \frac{p_i}{m}. \tag{3}$$

This simple, elegant set of equations is exactly what Laplace imagined the laws of nature to be: For each coordinate of phase space we have a single equation to tell you how it changes over an infinitesimal interval of time.

Action, Reaction, and the Conservation of Momentum

The principle of the conservation of momentum is a profound consequence of abstract general principles of classical mechanics that we have yet to formulate. But it can also be understood at an elementary level from *Newton's third law of motion*.

For every action there is an equal and opposite reaction.

The simplest way to think of the third law is to suppose first that particles interact in pairs. Each particle j exerts a force on each other particle i, and the total force on any particle is the sum of the forces on it exerted by all the other particles. If we denote the force on particle i due to particle j by the symbol $\vec{f}_{ij}$, then the total force acting on particle i is

$$\vec{F}_i = \sum_j \vec{f}_{ij}. \tag{4}$$

The left side represents the total force on particle i, and the right side is the sum of the forces acting on i due to all the other particles.

Newton's law of action and reaction is about the force between pairs of particles, $\vec{f}_{ij}$. What it says is simple: The force due to one particle j on another particle i is *equal and opposite* to the force due to particle i acting on particle j. As an equation, the third law says that for every i and j,

$$\vec{f}_{ij} = -\vec{f}_{ji}. \tag{5}$$

Let's rewrite the first of Equations (3), plugging in Eq. (4):

$$\dot{\vec{p}}_i = \sum_j \vec{f}_{ij}.$$

In other words, the rate of change of the momentum of any particle is the sum of the forces due to all the other particles. Now let's add up all these equations to see how the total momentum changes.

$$\sum_i \dot{\vec{p}}_i = \sum_i \sum_j \vec{f}_{ij}$$

The left-hand side of this equation is the sum of the rates of change of all the momenta (the plural of *momentum*). In other words, it is the rate of change of the total momentum. The right-hand side of the equation is zero. That's because when you write it out, each pair of particles contributes two terms: the force on i due to j, and the force on j due to i. The law of action and reaction, Eq. (5), ensures that these cancel. Thus we are left with an equation that we can write in the form

$$\frac{d}{dt} \sum_i \vec{p}_i = 0.$$

This equation is precisely the mathematical expression of the "conservation" of momentum: The total momentum of an isolated system never changes.

Let's consider the $6N$-dimensional space of p's and x's. At every point the entire collection of momenta are specified, so it follows that every point in the phase space is (partially) characterized by a value of the total momentum. We could go through the phase space, labeling each point with its total momentum. Now imagine starting the system of particles at some point. As time evolves, the phase point sweeps out a path in phase space. Every point on that path is labeled with the same value of total momentum; the point never jumps from one value to another. This is entirely similar to the idea of a conservation law that we explained in Lecture 1.

Lecture 5: Energy

"Old timer, what are you looking for under the locomotive?"

Lenny loved the big steam locomotives, so now and then, on their days off, George took him down to the train yard. Today, they found a confused old man who looked as if he had lost something.

"Where's the horse that pulls this thing?" the old timer asked George.

"We'll, it don't need no horse. Here, I'll show you how it works. You see this place over here," he said, pointing. "That's the fire box where they burn the coal to get out the chemical energy. Then this, right next to it, is the boiler where the heat boils the water to make steam. The steam pressure does work against the piston in this here box. Then, the piston pushes against these rods, and they make the wheels turn." The old timer grinned, shook George's hand, and took his leave.

Lenny had been standing aside while George explained the locomotive. Now, with a look of sheer admiration, he came over to George and said, "George, I loved the way you explained things to that guy. And I understood all of it. The fire box, the boiler, the piston. Just one thing I didn't get."

"What's that, Lenny?"

"Well, I was just wondering. Where's the horse?"

Force and Potential Energy

One often learns that there are many forms of energy (kinetic, potential, heat, chemical, nuclear, . . .) and that the sum total of all of them is conserved. But when reduced to the motion of particles, classical physics really has only two forms of energy: kinetic and potential. The best way to derive the conservation of energy is to jump right into the formal mathematical principles and then step back and see what we have.

The basic principle—call it the *potential energy principle*—asserts that all forces derive from a potential energy function denoted $V(\{x\})$. Recall that $\{x\}$ represents the entire set of $3N$ coordinates—the configuration space—of all particles in the system. To illustrate the principle, let's begin with the simplest case of a single particle moving along the x axis under the influence of a force $F(x)$. According to the potential energy principle, the force on the particle is related to the derivative of the potential energy, $V(x)$:

$$F(x) = -\frac{d\,V(x)}{d\,x}. \tag{1}$$

In the one-dimensional case, the potential energy principle is really just a definition of $V(x)$. In fact, the potential energy can be reconstructed from the force by integrating Eq. (1):

$$V(x) = -\int F(x)\,d\,x. \tag{2}$$

We can think of Eq. (1) in the following way: The force is always directed in a way that pushes the particle toward lower potential energy (note the minus sign). Moreover, the steeper $V(x)$, the stronger the force. The slogan that captures the idea is *Force pushes you down the hill.*

Potential energy by itself is not conserved. As the particle moves, $V(x)$ varies. What is conserved is the sum of potential energy and kinetic energy. Roughly speaking, as the particle rolls down the hill (in other words, as it moves toward lower potential energy), it picks up speed. As it rolls up the hill, it loses speed. Something is conserved.

Kinetic energy is defined in terms of the velocity v and mass m of the particle. It is denoted by T:

$$T = \frac{1}{2} m v^2.$$

The total energy E of the particle is the sum of the kinetic and potential energies:

$$E = \frac{1}{2} m v^2 + V(x).$$

As the particle rolls along the x axis, the two types of energy individually vary, but always in such a way that the sum is conserved. Let's prove it by showing that the time derivative of E is zero.

First let's calculate the rate of change of the kinetic energy. The mass is assumed constant, but v^2 can vary. The time derivative of v^2 is

$$\frac{d v^2}{d t} = 2 v \frac{d v}{d t} = 2 v \dot{v}. \tag{3}$$

Exercise 1: Prove Eq. (3). *Hint: Use the product rule for differentiation.*

It follows that the time derivative of the kinetic energy is

$$\dot{T} = m v \dot{v} = m v a,$$

where the time derivative of the velocity has been replaced by the acceleration.

Next, let's calculate the rate of change of potential energy. The key is to realize that $V(x)$ changes with time because x changes. Here is the formula that expresses this:

$$\frac{dV}{dt} = \frac{dV}{dx}\frac{dx}{dt}.$$

(It's okay to think of derivatives as ratios and to cancel the factors of dx in the numerator and denominator.) Another way to write this equation is to replace $\frac{dx}{dt}$ with the velocity v:

$$\frac{dV}{dt} = \frac{dV}{dx}v.$$

(Be careful to not confuse V and v.)

Now we can calculate the rate of change of the total energy:

$$\dot{E} = \dot{T} + \dot{V}$$

$$= mva + \frac{dV}{dx}v.$$

Note that since both terms contain a factor of v, we can factor it out:

$$\dot{E} = v\left(ma + \frac{dV}{dx}\right).$$

Now look at the expression in parentheses. Use the fact that the derivative of V is related to the force. Recalling the minus sign in Eq. (1), we see that the rate of change of E is given by

$$\dot{E} = v(m\,a - F(x))\,.$$

We now have what we need to prove energy conservation: Newton's law, $F = m\,a$, is exactly the condition that the factor in parentheses vanishes, which in turn tells us that the total energy is constant.

One point before we go on to many-dimensional motion. We have shown that energy is conserved, but why is it that momentum is not conserved in this case? After all, in the previous chapter we showed that for an isolated system of particles, Newton's third law implies that total momentum does not change. The answer is that we have left something out of the system—namely, the object that exerts the force on the one-dimensional particle. For example, if the problem has to do with a particle falling in a gravitational field, the gravitational force is exerted by the Earth. When the particle falls, its momentum changes, but that change is exactly compensated for by a tiny change in the motion of the Earth.

More Than One Dimension

It is a fact that the components of force are derivatives of potential energy, but it is not a definition. This is so when there is more than one x to worry about—because space has more than one dimension, or because there is more than one particle, or both. It is quite possible to imagine force laws that do not come from differentiating a potential energy function, but nature does not make use of such *nonconservative* forces.

Let's be a little more abstract than we have so far been. Call the coordinates of configuration space x_i (remember, configuration space is the same as position space). For now, the subscript i will not refer to which particle we are talking about or which direction of space. It runs over all these possibilities. In other words, for a system of N particles there are $3N$ values of i. Let's forget where they come from; we are simply considering a system of abstract coordinates labeled i.

Now let's write the equations of motion:

$$m_i \ddot{x}_i = F_i(\{x\}). \tag{4}$$

For each coordinate, there is a mass m_i and a component of force F_i. Each component of force can depend on all positions $\{x\}$.

We have seen in the one-dimensional case that the force is minus the derivative of the potential energy, as in Eq. (1). This was a definition of V, not a special condition on the force. But when there is more than one dimension, things get more complicated. It is generally *not* true that if you have a set of functions $F_i(\{x\})$, that they can all be derived by differentiating a single function $V(\{x\})$. It would be a brand-new principle if we asserted that the components of force can be described as (partial) derivatives of a single potential energy function.

Indeed this principle is not hypothetical. It is a basic mathematical expression of one of the most important principles of physics:

For any system there exists a potential $V(\{x\})$ such that

$$F_i(\{x\}) = -\frac{\partial V(\{x\})}{\partial x_i}. \tag{5}$$

What law of nature does Eq. (5) represent? You may have already guessed that it is the conservation of energy. We'll see that shortly, but first let's try to visualize what it means.

Picture a terrain with the function $V(\{x\})$ representing

the height or altitude at each point. First of all, the minus sign in Eq. (5) means that the force points in the downhill direction. It also says that the force is greater along directions where the slope is steeper. For example, on a contour map, there is no force pushing along the contour lines. The force vector points perpendicular to the contours.

Now let's come back and derive energy conservation. To do that, we plug Eq. (5) into the equations of motion (4):

$$m_i \ddot{x}_i = -\frac{\partial V(\{x\})}{\partial x_i}. \qquad (6)$$

The next step is to multiply each of the separate equations in Eq. (6) by the corresponding velocity $\dot{x}_i$ and sum them all,

$$\sum_i m_i \dot{x}_i \ddot{x}_i = -\sum_i \dot{x}_i \frac{\partial V(\{x\})}{\partial x_i}. \qquad (7)$$

Now we have to manipulate both sides of the equation in the same way that we did in the one-dimensional example. We define the kinetic energy to be the sum of all the kinetic energies for each coordinate:

$$T = \frac{1}{2} \sum_i m_i \dot{x}_i^2.$$

Here is what the two sides of Eq (7) give. First the left-hand side:

$$\sum_i m_i \dot{x}_i \ddot{x}_i = \frac{dT}{dt}.$$

Now the right-hand side:

$$-\sum_i \dot{x}_i \frac{\partial V(\{x\})}{\partial x_i} = -\frac{dV}{dt}.$$

Thus we can rewrite Eq. (7) as

$$\frac{dT}{dt} + \frac{dV}{dt} = 0. \tag{8}$$

Precisely as in the one-dimensional case, Eq. (8) says that the time derivative of the total energy is zero—energy is conserved.

To picture what is going on, imagine that the terrain has a frictionless ball rolling on it. Whenever the ball rolls toward a lower altitude it picks up speed, and whenever it rolls uphill it loses speed. The calculation tells us this happens in a special way that conserves the sum of the kinetic and potential energies.

You might wonder why the forces of nature are always gradients (derivatives) of a single function. In the next chapter we will reformulate classical mechanics using the principle of least action. In this formulation, it is "built in" from the very beginning that there is a potential energy function. But then why the principle of least action? Ultimately, the answer can be traced to the laws of quantum mechanics and to the origin of forces in field theory—subjects that, for the moment, are still out of range for us. So, why quantum field theory? At some point we have to give up and say that's just the way it is. Or, not give up and push on.

Exercise 2: Consider a particle in two dimensions, x and y. The particle has mass m, equal in both directions. The potential energy is $V = \frac{1}{2} k(x^2 + y^2)$. Work out the equations of motion. Show that there are circular orbits and that all orbits have the same period. Prove explicitly that the total energy is conserved.

Exercise 3: Rework Exercise 2 for the potential $V = \frac{k}{2(x^2+y^2)}$. Are there circular orbits? If so, do they all have the same period? Is the total energy conserved?

Before moving on to the principle of least action, I want to list a few of the different kinds of energy that we talk about in physics, and review how they fit into the picture. Let's consider

- mechanical energy
- heat
- chemical energy
- atomic/nuclear energy
- electrostatic energy
- magnetic energy
- radiation energy

Some, but not all, of these distinctions are a bit old-fashioned. *Mechanical energy* usually refers to the kinetic and potential energy of large visible objects such as planets or weights being hoisted by a crane. It often refers to gravitational potential energy.

The heat contained in a gas or other collection of molecules is also kinetic and potential energy. The only difference is that it involves the large and chaotic motion of so many particles that we don't even try to follow it in detail. Chemical energy is also a special case: The energy stored in chemical bonds is a combination of the potential energy and kinetic energy of the constituent particles that make up the molecules. It's harder to understand because quantum mechanics has to replace classical mechanics, but nonetheless, the energy is the potential and kinetic energy of particles. The same goes for atomic and nuclear energy.

Electrostatic energy is just another word for the potential energy associated with the forces of attraction and repulsion between electrically charged particles. In fact, apart from gravitational energy, it is the primary form of potential energy in the ordinary, classical world. It is the potential energy between charged particles in atoms and molecules.

Magnetic energy is tricky, but the force between the poles of magnets is a form of potential energy. The tricky part comes when we think about the forces between magnets and charged particles. Magnetic forces on charged particles are a new kind of beast called velocity-dependent forces. We will come back to this later in the book.

Finally, there is the energy stored in electromagnetic radiation. It can take the form of heat from the sun, or the energy stored in radio waves, laser light, or other forms of radiation. In some very general sense, it is a combination of kinetic and potential energy, but it is not the energy of particles (not until we get to quantum field theory, anyway) but, of fields. So we will set electromagnetic energy aside until a later book.

Lecture 6: The Principle of Least Action

Lenny was frustrated—not a good sign considering his size and strength—and his head hurt. "George, I can't remember all this stuff! Forces, masses, Newton's equations, momentum, energy. You told me that I didn't need to memorize stuff to do physics. Can't you make it just one thing to remember?"

"Okay, Lenny. Calm down. I'll make it simple. All you have to remember is that the action is always stationary."

The Transition to Advanced Mechanics

The principle of least action—really the principle of stationary action—is the most compact form of the classical laws of physics. This simple rule (it can be written in a single line) summarizes everything! Not only the principles of classical mechanics, but electromagnetism, general relativity, quantum mechanics, everything known about chemistry—right down to the ultimate known constituents of matter, elementary particles.

Let's begin with a general observation about the basic problem of classical mechanics, namely this problem is to determine the trajectories (or orbits) of systems from their equations of motion. We usually express the problem by postulating three things: the masses of the particles, a set of forces $F(\{x\})$ (or, even better, a formula for the potential energy), and an initial condition. The system begins with some values of the coordinates and velocities and then moves, according to Newton's second law, under the influence of the given forces. If there are a total of N coordinates, $(x_1, x_2, ..., x_N)$, then the initial conditions consist of specifying the $2N$ positions and velocities. For example, at an initial time t_0, we can specify the positions $\{x\}$ and velocities $\{\dot{x}\}$ and then solve the equations to find out what the positions and velocities will be at the later time t_1. In the process, we will usually determine the whole trajectory between t_0 and t_1 (see Figure 1).

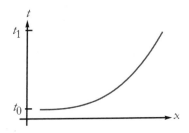

Figure 1: A trajectory from time t_0 to time t_1.

But we can formulate the problem of classical mechanics in another way that also involves specifying $2N$ items of information. Instead of providing the initial positions and velocities, we provide the initial and final positions. Here is the way to think about it: Suppose that an outfielder wants to throw a baseball (from x_0 at time t_0) and he wants it to arrive at second base (x_1) after exactly 1.5 seconds (t_1). How does the ball have to move in between? Part of the problem in this case will be to determine what the initial velocity of the ball has to be. The initial velocity is not part of the input data in this way of posing the question; it is part of the solution.

Let's draw a space-time picture to illustrate the point (see Figure 2). The horizontal axis shows the position of a particle (or the baseball), and the vertical axis denotes the time. The beginning and end of the trajectory are a pair of points on the space-time diagram, and the trajectory itself is a curve connecting the points.

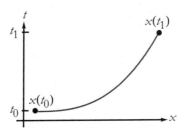

Figure 2: A trajectory of the baseball.

The two ways of posing the problem of motion are analogous to two ways of formulating the problem of fixing a straight line in space. One thing we could ask is to construct a straight line from the origin that begins in some particular direction. That's like asking for the trajectory given the initial position and velocity. On the other hand, we could ask to construct a straight line that connects two particular points. That is like finding the trajectory that begins at one position and arrives at another position after a specified time. In this form, the problem is similar to asking how we have to aim a line from some initial point so that it passes through another point. The answer: Find the shortest path between the points. In the problems of classical mechanics, the answer is to find the path of stationary action.

Action and the Lagrangian

Formulating the action principle involves exactly the same parameters as formulating Newton's equations. You have to know the masses of the particles, and you have to know the potential energy. The action for a trajectory is an integral from the start of the trajectory at t_0 to the end of the trajectory at t_1. I'll just tell you what the integral is—no motivation—and then we'll explore the consequences of minimizing it.[1] We'll end up with Newton's equations. Once we see how that works, any further motivation will be unnecessary. If it's equivalent to Newton's equations, what more motivation do we need?

Before being general, let's illustrate the idea for a single particle moving on a line. The position of the particle at time t is $x(t)$, and its velocity is $\dot{x}(t)$. The kinetic and potential energies are

$$T = \frac{1}{2} m \dot{x}^2$$
$$V = V(x),$$

respectively. The action of a trajectory is written

$$\mathcal{A} = \int_{t_0}^{t_1} (T - V)\, d t$$
$$= \int_{t_0}^{t_1} \left(\frac{1}{2} m \dot{x}^2 - V(x) \right) d t. \tag{1}$$

You might think that there is a typo in Eq. (1). The energy is the sum of T and V, but the integral involves the difference. Why the difference and not the sum? You can try the derivation with $T + V$, but you'll get the wrong answer. The quantity $T - V$ is

---

[1.] I use the term minimizing because, to my knowledge, there is no verb to express making a quantity stationary. I tried stationaryizing, stationizing, and a few others, but I eventually gave up and took the path of least action. But remember, least action really means stationary action.

called the *Lagrangian* of the system, and it's denoted by the symbol L. The things you need to know to specify L are the mass of the particle (for the kinetic energy) and the potential $V(x)$. It is, of course, no accident that these are the same things you need to know to write Newton's equation of motion.

Think of the Lagrangian as a function of the position x and the velocity $\dot{x}$. It's a function of position because the potential energy depends on x, and it's a function of velocity because the kinetic energy depends on $\dot{x}$. So we write

$$L = L(x, \dot{x}).$$

We can rewrite the action as the integral of the Lagrangian:

$$\mathcal{A} = \int_{t_0}^{t_1} L(x, \dot{x}) \, d t. \qquad (2)$$

The principle of stationary action is really very remarkable. It almost seems that the particle must have supernatural powers to feel out all the possible trajectories and pick the one that makes the action stationary. Let's pause to consider what we are doing and where we are going.

The process of minimizing the action is a generalization of minimizing a function. The action is not an ordinary function of a few variables. It depends on an infinity of variables: All the coordinates *at every instant of time.* Imagine replacing the continuous trajectory by a "stroboscopic" trajectory consisting of a million points. Each point is specified by a coordinate x, but the whole trajectory is specified only when a million x's are specified. The action is a function of the whole trajectory, so it is a function of a million variables. Minimizing the action involves a million equations.

Time is not really stroboscopic, and a real trajectory is a function of a continuously infinite number of variables. To put it another way, the trajectory is specified by a function $x(t)$, and the action is a function of a function. A function of a function—a quantity that depends on an entire function—is called a *functional*. Minimizing a functional is the subject of a branch of mathematics called the *calculus of variations.*

Nevertheless, despite the differences from ordinary functions, the condition for a stationary action strongly resembles the condition for a stationary point of a function. In fact, it has exactly the same form as Eq. (4) in Interlude 3, namely

$$\delta A = 0.$$

Now, however, the variations are not just small shifts of a few coordinates, but all the possible small variations of a whole trajectory.

Later in this lecture we will work out the equations for minimizing the action. They are called the Euler-Lagrange equations. For the case of a single degree of freedom, there is one equation at each point along the trajectory. In fact, the equations become differential equations that tell the system how to move from one instant to the next. Thus the particle does not have to have supernatural powers to test out all future trajectories—at least no more so than it needs to follow Newton's equations of motion.

We will derive the Euler-Lagrange equations later in this lecture. To do you a flavor, I will write down their form. If you are the independent type, you can try to plug in the Lagrangian and see if you can derive Newton's equation of motion. Here, then, is the Euler-Lagrange equation for a single degree of freedom:

$$\frac{d}{dt}\frac{\partial L}{\partial \dot{x}} - \frac{\partial L}{\partial x} = 0.$$

Derivation of the Euler-Lagrange Equation

Let's see if we can derive the Euler-Lagrange equation for a single degree of freedom. Start by replacing continuous time with stroboscopic time. The instants can be labeled by integers n. The time between neighboring instants is very small. Call it Δt. The action is an integral, but, as always, an integral is the limit of a sum. In this case, we are going to think of the sum as being over the intervals between successive instants.

Here are the replacements that we do when we approximate the integral by the sum:

$$\int L\,dt = \sum L\,\Delta t$$

$$\dot{x} = \frac{x_{n+1} - x_n}{\Delta t}.$$

The first replacement is just the usual approximation of replacing the integral by a discrete sum of terms, each weighted with the small time interval Δt. The second is also familiar. It replaces the velocity $\dot{x}$ with the difference of neighboring positions divided by the small time interval.

The last replacement is a bit more subtle. Since we are

going to think of the sum as being over the small intervals between neighboring instants, we need an expression for the position halfway between the instants. That's easy. Just replace $x(t)$ with the average position between neighboring instants:

$$x(t) = \frac{x_n + x_{n+1}}{2}.$$

Notice that everywhere $\dot{x}$ occurred in the Lagrangian I replaced it with $\frac{x_{n+1} - x_n}{\Delta t}$, and everywhere that x occurred I substituted $\frac{x_n + x_{n+1}}{2}$.

The total action is found by adding up all the incremental contributions:

$$A = \sum_n L\left(\frac{x_{n+1} - x_n}{\Delta t}, \frac{x_n + x_{n+1}}{2}\right)\Delta t. \qquad (3)$$

I have very explicitly taken the action apart into its components, almost like writing a computer program to evaluate it.

Now suppose we want to minimize the action by varying any one of the x_n and setting the result equal to zero. Let's pick one of them, say x_8. (Any other one would have been just as good.) This sounds very complicated, but notice that x_8 appears only in two of the terms in Eq. (3). The two terms that contain x_8 are

$$A = L\left(\frac{x_9 - x_8}{\Delta t}, \frac{x_8 + x_9}{2}\right)\Delta t +$$
$$L\left(\frac{x_8 - x_7}{\Delta t}, \frac{x_7 + x_8}{2}\right)\Delta t.$$

Now all we have to do is differentiate with respect to x_8. Notice that x_8 appears in two ways in each term. It appears through the velocity dependence and through the x dependence. The derivative of A with respect to x_8 is

$$\frac{\partial A}{\partial x_8} = \frac{1}{\Delta t}\left(-\frac{\partial L}{\partial \dot{x}}\bigg|_{n=9} + \frac{\partial L}{\partial \dot{x}}\bigg|_{n=8}\right) +$$
$$\frac{1}{2}\left(\frac{\partial L}{\partial x}\bigg|_{n=8} + \frac{\partial L}{\partial x}\bigg|_{n=9}\right).$$

The symbol $|_{n=8}$ is an instruction to evaluate the function at the discrete time $n = 8$.

To minimize the action with respect to variations of x_8, we set dA/dx equal to zero. But before we do, let's see what happens to dA/dx in the limit when Δt tends to zero. Start with the first term,

$$\frac{1}{\Delta t}\left(-\frac{\partial L}{\partial \dot{x}}\bigg|_{n=9} + \frac{\partial L}{\partial \dot{x}}\bigg|_{n=8}\right).$$

This has the form of the difference between a quantity evaluated at two neighboring times, $n = 8$ and $n = 9$, divided by the small separation between them. This obviously tends to a derivative, namely

$$\frac{1}{\Delta t}\left(-\frac{\partial L}{\partial \dot{x}}\bigg|_{n=9} + \frac{\partial L}{\partial \dot{x}}\bigg|_{n=8}\right) \longrightarrow -\frac{d}{dt}\frac{\partial L}{\partial \dot{x}}.$$

The second term,

$$\frac{1}{2}\left(\frac{\partial L}{\partial x}\bigg|_{n=8} + \frac{\partial L}{\partial x}\bigg|_{n=9}\right),$$

also has a simple limit. It is half the sum of $\frac{\partial L}{\partial x}$ evaluated at neighboring times. As the separation between the points shrinks to zero, we just get $\dfrac{\partial L}{\partial x}$.

The condition that $\dfrac{\partial A}{\partial x_8} = 0$ becomes the Euler-Lagrange equation,

$$\frac{d}{dt} \frac{\partial L}{\partial \dot{x}} - \frac{\partial L}{\partial x} = 0. \tag{4}$$

Exercise 1: Show that Eq. (4) is just another form of Newton's equation of motion $F = m\,a$.

The derivation is essentially the same for many degrees of freedom. There is an Euler-Lagrange equation for each coordinate x_i:

$$\frac{d}{dt} \frac{\partial L}{\partial \dot{x}_i} - \frac{\partial L}{\partial x_i} = 0.$$

What this derivation shows is that there is no magic involved in the ability of the particle to feel out the entire path before deciding which way to go. At each stage along the trajectory, the particle has only to minimize the action between a point in time and a neighboring point in time. The principle of least action just becomes a differential equation at each instant that determines the immediate future.

More Particles and More Dimensions

All together, let there be N coordinates that we call x_i. The motion of the system is described by a trajectory, or *orbit*, through an N-dimensional space. For an even better description we can add time, thinking of the orbit as a path through $N + 1$

dimensions. The starting point of the trajectory is the set of points $x_i(t_0)$, and the endpoint is another set of points $x_i(t_1)$. The orbit through the $(N + 1)$-dimensional space is described by giving all the coordinates as functions of time $x_i(t)$.

The principle of least action for more degrees of freedom is essentially no different than the case with only a single degree of freedom. The Lagrangian is the kinetic energy minus the potential energy:

$$L = \sum_i \left(\frac{1}{2} m_i \dot{x}_i{}^2 \right) - V(\{x\}).$$

The action is also exactly as before, the integral of the Lagrangian,

$$\mathcal{A} = \int_{t_0}^{t_1} L\left(\{x\}, \{\dot{x}\} \right) d\, t, \tag{5}$$

and the principle of least (stationary) action is that the trajectory minimizes this action.

When there are many variables, we can vary the trajectory in many ways, for example we can vary $x_1(t)$, or $x_2(t)$, and so on. It's like minimizing a function of many variables: There is an equation for each variable. The same is true for the Euler-Lagrange equations: There is one for each variable x_i. Each one has the same general form as Eq. (4)

$$\frac{d}{d\, t} \left(\frac{\partial L}{\partial \dot{x}_i} \right) = \frac{\partial L}{\partial x_i}. \tag{6}$$

Exercise 2: Show that Eq. (6) is just another form of Newton's equation of motion $F_i = m_i \ddot{x}_i$.

What's Good about Least Action?

There are two primary reasons for using the principle of least action. First, it packages everything about a system in a very concise way. All the parameters (such as the masses and forces), and all the equations of motion are packaged in a single function—the Lagrangian. Once you know the Lagrangian, the only thing left to specify is the initial conditions. That's really an advance: a single function summarizing the behavior of any number of degrees of freedom. In future volumes, we will find that whole theories—Maxwell's theory of electrodynamics, Einstein's theory of gravity, the Standard Model of elementary particles—are each described by a Lagrangian.

The second reason for using the principle of least action is the practical advantage of the Lagrangian formulation of mechanics. We'll illustrate it by an example. Suppose we want to write Newton's equations in some other coordinates, or in some frame of reference that is moving or accelerating.

Take the case of a particle in one dimension that, from the point of view of someone standing at rest, satisfies Newton's laws. The physicist at rest—call him Lenny—uses the coordinate x to locate the object.

A second physicist—George—is moving relative to Lenny, and he wants to know how to describe the object relative to his own coordinates. First of all, what does it mean to talk about George's coordinates? Because George moves relative to Lenny, the origin of his coordinate frame moves relative to Lenny's origin. This is easily described by changing coordinates from Lenny's x to George's coordinate system X.

Here is how we do it. At any time t, Lenny locates George's origin at $x + f(t)$, where f is some function that describes how George moves relative to Lenny. An event (at time *t*) that Lenny assigns a coordinate x, George assigns coordinate X where

$$X = x - f(t).$$

When Lenny sees a particle moving on the trajectory $x(t)$, George sees the same particle moving on the trajectory $X = x(t) - f(t)$. If George does not want to keep asking Lenny what the trajectory is, then he wants his own laws of motion to describe the object from his coordinates. The easiest way to do this is to *transform* the equations of motion from one coordinate system to another is to use the principle of least action, or the Euler-Lagrange equations.

According to Lenny, the action of a trajectory is

$$\mathcal{A} = \int_{t_0}^{t_1} \left(\frac{1}{2} m \dot{x}^2 - V(x) \right) dt. \tag{7}$$

But we can also write the action in terms of George's coordinates. All we have to do is to express $\dot{x}$ in terms of $\dot{X}$:

$$\dot{x} = \dot{X} + \dot{f}.$$

So we plug this into Eq. (7) to get

$$\mathcal{A} = \int_{t_0}^{t_1} \frac{1}{2} m \left(\dot{X} + \dot{f} \right)^2 - V(X) \, dt.$$

The potential energy $V(X)$ simply means the potential energy that Lenny would use, evaluated at the object's location, but expressed in George's coordinates—same point, different label. And now we know the Lagrangian in the X frame of reference,

$$L = \frac{1}{2}m\left(\dot{X} + \dot{f}\right)^2 - V(X),$$

where we can expand the square:

$$L = \frac{1}{2}m\left(\dot{X}^2 + 2\dot{X}\dot{f} + \dot{f}^2\right) - V(X). \tag{8}$$

What does George do with Eq. (8)? He writes the Euler-Lagrange equation. Here is what he gets:

$$m\ddot{X} + m\ddot{f} = -\frac{dV}{dX},$$

or, with a small rearrangement,

$$m\ddot{X} = -\frac{dV}{dX} - m\ddot{f}.$$

The result is nothing surprising. George sees an extra, "fictitious" force on the object equal to $-m\ddot{f}$. What is interesting is the procedure: Instead of transforming the equation of motion, we worked directly with the Lagrangian.

Let's do another example. This time George is on a rotating carousel. Lenny's coordinates are x and y. George's coordinate frame is X and Y, and it rotates with the carousel. Here is the connection between the two frames:

$$\begin{aligned} x &= X\cos\omega t + Y\sin\omega t \\ y &= -X\sin\omega t + Y\cos\omega t. \end{aligned} \tag{9}$$

Both observers see a particle moving in the plane. Let's assume that Lenny observes that the particle moves with no forces acting on it. He describes the motion using the action principle with Lagrangian

$$L = \frac{m}{2}\left(\dot{x}^2 + \dot{y}^2\right). \tag{10}$$

What we want to do is express the action in George's rotating frame and then use the Euler-Lagrange equations to figure out the equations of motion. Since we already know the action in Lenny's frame, all we need to do is express the velocity in his frame in terms of George's variables. Just differentiate Equations (9) with respect to time:

$$\dot{x} = \dot{X}\cos\omega t - \omega X \sin\omega t + \dot{Y}\sin\omega t$$
$$+ \omega Y \cos\omega t$$
$$\dot{y} = -\dot{X}\sin\omega t - \omega X \cos\omega t + \dot{Y}\cos\omega t$$
$$- \omega Y \sin\omega t.$$

After a little bit of algebra using $\sin^2 + \cos^2 = 1$, here is what we get for $\dot{x}^2 + \dot{y}^2$

$$\dot{x}^2 + \dot{y}^2 = \dot{X}^2 + \dot{Y}^2 + \omega^2\left(X^2 + Y^2\right) + 2\omega\left(\dot{X}Y - \dot{Y}X\right). \tag{11}$$

Now all we have to do is plug Eq. (11) into Lenny's Lagrangian, Eq. (10), to get George's Lagrangian. It's the same Lagrangian except expressed in George's coordinates:

$$L = \frac{m}{2}\left(\dot{X}^2 + \dot{Y}^2\right) + \frac{m\,\omega^2}{2}\left(X^2 + Y^2\right) + $$
$$m\,\omega\left(\dot{X}\,Y - \dot{Y}\,X\right). \tag{12}$$

Let's examine the various terms. The first term, $\frac{m}{2}\left(\dot{X}^2 + \dot{Y}^2\right)$, is familiar—it's just what George would call the kinetic energy. Notice that if angular velocity were zero, that's all there would be. The next term, $m\omega^2\left(X^2 + Y^2\right)$, is something new due to the rotation. What it looks like to George is a potential energy,

$$V = -m\,\omega^2\left(X^2 + Y^2\right),$$

which can easily be seen to create an outward force proportional to the distance from the center of rotation:

$$F = m\,\omega^2\,\vec{r}.$$

This is nothing but the centrifugal force.

The last term in Eq. (12) is a little less familiar. It is called the *Coriolis force*. To see how it all works, we can work out the Euler-Lagrange equations. Here is what we get:

$$m\,\ddot{X} = m\,\omega^2\,X - 2\,m\,\omega\,\dot{Y}$$
$$m\,\ddot{Y} = m\,\omega^2\,Y + 2\,m\,\omega\,\dot{X}.$$

This looks exactly like Newton's equations with centrifugal and Coriolis forces. Notice that there is something new in the form of the force law. The components of the Coriolis force,

$$F_X = -2\, m\, \omega\, \dot{Y}$$

$$F_Y = 2\, m\, \omega\, \dot{X},$$

depend not only on the position of the particle but also on its velocity. The Coriolis force is a velocity-dependent force.

Exercise 3: Use the Euler-Lagrange equations to derive the equations of motion from this Lagrangian.

The main point of this exercise was not so much to derive the centrifugal and Coriolis forces as to show you how to transform a mechanics problem from one coordinate system to another by simply rewriting the Lagrangian in the new coordinates. This is, by far, the easiest way to do the transformation—a lot easier than trying to transform Newton's equations directly.

Another example, which we will leave to you, is to transform George's equations to polar coordinates:

$$X = R \cos \theta$$
$$Y = R \sin \theta.$$

Exercise 4: Work out George's Lagrangian and Euler-Lagrange equations in polar coordinates.

Generalized Coordinates and Momenta

There is really nothing very general about Cartesian coordinates. There are many coordinate systems that we can choose to represent any mechanical system. For example, suppose we want to study the motion of an object moving on a spherical surface—

say, the Earth's surface. In this case, Cartesian coordinates are not of much use: The natural coordinates are two angles, longitude and latitude. Even more general would be an object rolling on a general curved surface like a hilly terrain. In such a case, there may not be any special set of coordinates. That's why it is important to set up the equations of classical mechanics in a general way that applies to any coordinate system.

Consider an abstract problem in which a system is specified by a general set of coordinates. We usually reserve the notations x_i for Cartesian coordinates. The notation for a general system of coordinates is called q_i. The q_i could be Cartesian coordinates, or polar coordinates, or anything else we can think of.

We also need to specify the velocities, which in the abstract situation means the time derivatives of the q_i generalized coordinates. An initial condition consists of the set of generalized coordinates and velocities $\left(q_i, \dot{q}_i\right)$.

In a general coordinate system, the equations of motion may be complicated, but the action principle always applies. All systems of classical physics—even waves and fields—are described by a Lagrangian. Sometimes the Lagrangian is calculated from some previous knowledge. An example is calculating George's Lagrangian, knowing Lenny's. Sometimes the Lagrangian is guessed on the basis of some theoretical prejudices or principles, and sometimes we deduce it from experiments. But however we get it, the Lagrangian neatly summarizes all the equations of motion in a simple package.

Why are all systems described by action principles and Lagrangians? It's not easy to say, but the reason is very closely related to the quantum origins of classical physics. It is also closely related to the conservation of energy. For now, we are

$$p_i = \frac{\partial L}{\partial \dot{q}_i}.$$

The notation for generalized momentum is p_i.

With that definition, the Euler-Lagrange equations are

$$\frac{d\,p_i}{d\,t} = \frac{\partial L}{\partial q_i}.$$

Let's do a couple of examples starting with a particle in polar coordinates. In this case the q_i's are the radius, r, and the angle, θ. We can use the result from Exercise 4 to get the Lagrangian:

$$L = \frac{m}{2}\left(\dot{r}^2 + r^2\,\dot{\theta}^2\right).$$

The generalized momentum conjugate to r (the r momentum) is

$$p_r = \frac{\partial L}{\partial \dot{r}} = m\,\dot{r},$$

and the corresponding equation of motion is

$$\frac{d\,p_r}{d\,t} = \frac{\partial L}{\partial r} = m\,r\,\dot{\theta}^2.$$

Using $\dot{p} = m\,\ddot{r}$ and canceling m from both sides, we can write this equation in the form

$$\ddot{r} = r\,\dot{\theta}^2.$$

The equation of motion for the angle θ is especially interesting. First consider the conjugate momentum to θ:

going to it take as given that all known systems of classical physics can be described in terms of the action principle.

The Lagrangian is always a function of the coordinates and the velocities, $L = L(q_i, \dot{q}_i)$, and the action principle is always

$$\delta \mathcal{A} = \delta \int_{t_0}^{t_1} L(q_i, \dot{q}_i) \, d\,t = 0.$$

This means that the equations are of the Euler-Lagrange form. Here, then, is the most general form of classical equations of motion. There is an equation for each q_i

$$\frac{d}{d\,t} \left(\frac{\partial L}{\partial \dot{q}_i} \right) = \frac{\partial L}{\partial q_i}. \tag{13}$$

That's it, all of classical physics in a nutshell! If you know what the q_i's are, and if you know the Lagrangian, then you have it all.

Let's look a little closer at the two sides of Eq. (13). Begin with the expression $\dfrac{\partial L}{\partial \dot{q}_i}$. Suppose for a moment that the q_i's are the ordinary Cartesian coordinates of a particle and L is the usual kinetic energy minus potential energy. In this case, the Lagrangian would contain $\dfrac{m}{2} \dot{x}^2$ and then $\dfrac{\partial L}{\partial \dot{q}_i}$ would just be $m\dot{x}$—in other words, the component of momentum in the x direction. We then call $\dfrac{\partial L}{\partial \dot{q}_i}$ the *generalized momentum conjugate* to q_i or just the *conjugate momentum* to q_i.

The concept of conjugate momentum transcends the simple example in which momentum comes out to be mass times velocity. Depending on the Lagrangian, the conjugate momentum may not be anything you recognize, but it is always defined by

$$p_\theta = \frac{\partial L}{\partial \dot\theta} = m\, r^2\, \dot\theta.$$

This quantity should be familiar. It is the *angular momentum* of the particle. Angular momentum and p_θ are exactly the same thing.

Now consider the equation of motion for θ. Since θ itself does not appear in the Lagrangian, there is no right-hand side, and we have

$$\frac{d\, p_\theta}{d\, t} = 0. \tag{14}$$

In other words, angular momentum is conserved. Another way to write Eq. (14) is

$$\frac{d}{d\, t}\left(m\, r^2\, \dot\theta\right) = 0. \tag{15}$$

We can see that $r^2\, \dot\theta$ is a constant. That's why angular velocity increases as a particle gets closer to the origin.

Exercise 5: Use these results to predict the motion of a pendulum of length *l*.

Cyclic Coordinates

As we've just seen, it sometimes happens that some coordinate does not appear in the Lagrangian though its velocity does. Such coordinates are called *cyclic* (I don't know why.)

What we *do* know is that the Lagrangian does not change when you shift the value of a cyclic coordinate. Whenever a coordinate is cyclic, its conjugate momentum is conserved.

Angular momentum is one example. Another is ordinary (linear) momentum. Take the case of a single particle with Lagrangian

$$L = \frac{m}{2}\left(\dot{x}^2 + \dot{y}^2 + \dot{z}^2\right).$$

None of the coordinates appear in the Lagrangian, so they are all cyclic. Again, there is nothing particularly cyclic about them—it's just a word. Therefore, all of the components of momentum are conserved. This would not be true if there were a potential energy that depended on the coordinates.

Let's take another case: two particles moving on a line with a potential energy that depends on the distance between them. For simplicity I'll take the masses to be equal, but there is nothing special about that case. Let's call the positions of the particles x_1 and x_2. The Lagrangian is

$$L = \frac{m}{2}\left(\dot{x_1}^2 + \dot{x_2}^2\right) - V(x_1 - x_2). \tag{16}$$

Now the Lagrangian depends on both x_1 and x_2, and neither is cyclic. Neither momentum is conserved.

But that's missing an important point. Let's make a change of coordinates. Define x_+ and x_- as

$$x_+ = \frac{(x_1 + x_2)}{2}$$

$$x_- = \frac{(x_1 - x_2)}{2}.$$

We can easily rewrite the Lagrangian. The kinetic energy is

$$T = m\left(\dot{x_+}^2 + \dot{x_-}^2\right).$$

Exercise 6: Explain how we derived this.

The important point is that the potential energy depends only on x_-. The Lagrangian is then

$$L = m\left(\dot{x}_+^2 + \dot{x}_-^2\right) - V(x_-).$$

In other words, there was a hidden cyclic coordinate, namely x_+. This means that the conjugate momentum to x_+ (call it p_+) is conserved. It is easy to see that p_+ is nothing but the total momentum,

$$p_+ = 2m\dot{x}_+ = m\dot{x}_1 + m\dot{x}_2.$$

The real point that we will come to in the next lecture is not so much about cyclic coordinates but about symmetries.

Lecture 7: Symmetries and Conservation Laws

Lenny had trouble reading maps. It always seemed like whichever way he was facing must be north. He wondered why he had more trouble with NSEW than he did with up and down. He could almost always get up and down right.

Preliminaries

The relationship between symmetries and conservation laws is one of the big main themes of modern physics. We're going to begin by giving some examples of conservation laws for some simple systems. At first, the fact that certain quantities are conserved will seem somewhat accidental—hardly things of deep principle. Our real goal, however, is not to identify accidental conserved quantities, but to identify a set of principles connecting them to something deeper.

We'll begin with the system that we studied at the end of Lecture 6 in Eq. (16), but let's free it from the interpretation of particles moving on a line. It could be any system with two coordinates: particles, fields, rotating rigid bodies, or whatever. To emphasize the broader context, let's call the coordinates q instead of x and write a Lagrangian of similar—but not quite identical—form:

$$L = \frac{1}{2}\left(\dot{q_1}^2 + \dot{q_2}^2\right) - V(q_1 - q_2). \tag{1}$$

The potential is a function of one combination of variables, namely $(q_1 - q_2)$. Let's denote the derivative of the potential V by V'. Here are the equations of motion:

$$\dot{p}_1 = -V'(q_1 - q_2)$$
$$\dot{p}_2 = +V'(q_1 - q_2). \tag{2}$$

Exercise 1: Derive Equations (2) and explain the sign difference.

Now add the two equations together to see that the sum $p_1 + p_2$ is conserved.

Next, let's do something slightly more complicated. Instead of the potential being a function of $(q_1 - q_2)$, let's have it be a function of a general linear combination of q_1 and q_2. Call the combination $(a\,q_1 - b\,q_2)$. The potential then has the form

$$V(q_1, q_2) = V(a\,q_1 - b\,q_2). \tag{3}$$

For this case the equations of motion are

$$\dot{p}_1 = -a\,V'(a\,q_1 - b\,q_2)$$
$$\dot{p}_2 = +b\,V'(a\,q_1 - b\,q_2).$$

It seems that we've lost the conservation law; adding the two equations does not give the conservation of $p_1 + p_2$.

But the conservation law has not been lost; it just changed a little bit. By multiplying the first equation by b and the second by a and then adding them, we can see that $b\,p_1 + a\,p_2$ is conserved.

Exercise 2: Explain this conservation.

On the other hand, suppose the potential is a function of some other, more general combination of the q's, such as $q_1 + q_2^2$. Then there is no conserved combination of the p's. So, then, what is the principle? What determines whether there are conservation laws and what they are? The answer has been known for almost 100 years from the work of the German mathematician Emmy Noether.

Examples of Symmetries

Let's consider a change of coordinates from q_i to a new set q_i'. Each q_i' is a function of all of the original q coordinates:

$$q_i' = q_i'(q_i).$$

There are two ways to think about a change of coordinates. The first way is called *passive*. You don't do anything to the system—just relabel the points of the configuration space.

For example, suppose that the x axis is labeled with tick marks, $x = \ldots, -1, 0, 1, 2, \ldots$ and there is a particle at $x = 1$. Now suppose you are told to perform the coordinate transformation

$$x' = x + 1. \tag{4}$$

According to the passive way of thinking, the transformation consists of erasing all the labels and replacing them with new ones. The point formerly known as $x = 0$ is now called $x' = 1$. The point formerly known as $x = 1$ is now called $x' = 2$, and so on. But the particle is left where it was (if it was at $x = 1$, then the new labeling puts it at $x' = 2$); only the label has changed.

In the second way of thinking about coordinate transformations, which is called *active*, you don't relabel the points at all. The transformation $x' = x + 1$ is interpreted as an instruction: Wherever the particle is, move it one unit to the right. In other words, it is an instruction to actually move the system to a new point in the configuration space.

In what follows, we will adopt the active point of view. Whenever I make a change of coordinates, it means that the system is actually displaced to the new point in the configuration space. In general, when we make a transformation, the system actually changes. If, for example, we move an object, the potential energy—and therefore the Lagrangian—may change.

Now I can explain what a symmetry means. A *symmetry* is an active coordinate transformation that does not change the value of the Lagrangian. In fact, no matter where the system is located in the configuration space, such a transformation does not change the Lagrangian.

Let's take the simplest example: a single degree of freedom with Lagrangian

$$L = \frac{1}{2}\dot{q}^2.$$

Suppose we make a change in the coordinate q by shifting it an amount δ. In other words, any configuration is replaced by another in which q has been shifted (see Figure 1).

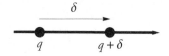

Figure 1: Shifting the coordinate of a point, q, by δ.

If the shift δ does not depend on time (as we will assume), then the velocity $\dot{q}$ does not change, and—most important—neither does the Lagrangian. In other words, under the change

$$q \rightarrow q + \delta, \tag{5}$$

the change in the Lagrangian is $\delta L = 0$.

In Eq. (5) the quantity δ can be any number. Later, when we consider transformations by infinitesimal steps, the symbol δ will be used to represent infinitesimal quantities, but for now it doesn't matter.

We could consider a more complicated Lagrangian with a potential energy $V(q)$. Unless the potential is a constant independent of q, then the Lagrangian will change as q is shifted. In that case there is no symmetry. The symmetry of moving a system in space by adding a constant to the coordinates is called *translation symmetry*, and we will spend a lot of time discussing it.

Now look at Equations (2). Suppose we shift q_1 but not q_2. In that case the Lagrangian will change because the potential energy changes. But if we shift both coordinates by the same amount so that $q_1 - q_2$ does not change, then the value of the Lagrangian is unchanged. We say that the Lagrangian is *invariant* under the change

$$\begin{aligned} q_1 &\rightarrow q_1 + \delta \\ q_2 &\rightarrow q_2 + \delta. \end{aligned} \tag{6}$$

We say that the Lagrangian is symmetric with respect to the transformation in Equations (6). Again this is a case of translation symmetry, but in this case, to have a symmetry we must translate both particles so that the distance between them is unchanged.

For the more complicated case of Eq. (3), where the potential depends on $a\,q_1 + b\,q_2$, the symmetry is less obvious. Here is the transformation:

$$q_1 \rightarrow q_1 + b\,\delta$$
$$q_2 \rightarrow q_2 - a\,\delta. \tag{7}$$

> **Exercise 3: Show that the combination $a\,q_1 + b\,q_2$, along with the Lagrangian, is invariant under Equations (7).**

If the potential is a function of a more complicated combination, it is not always clear that there will be a symmetry. To illustrate a more complex symmetry, let's revert to Cartesian coordinates for a particle moving on the x, y plane. Let's say this particle is under the influence of a potential energy that depends only on the distance from the origin:

$$L = \frac{m}{2}\left(\dot{x}^2 + \dot{y}^2\right) - V\left(x^2 + y^2\right). \tag{8}$$

It's pretty obvious that Eq. (8) has a symmetry. Imagine rotating the configuration about the origin by an angle θ (see Figure 2).

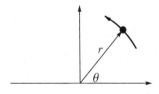

Figure 2: Rotation by θ.

Since the potential is a function only of the distance from the origin, it doesn't change if the system is rotated through an angle.

Moreover, the kinetic energy is also unchanged by a rotation. The question is how we express such a change. The answer is obvious: Just rotate coordinates

$$x \rightarrow x\cos\theta + y\sin\theta$$
$$y \rightarrow -x\sin\theta + y\cos\theta. \tag{9}$$

where θ is any angle.

Now we come to an essential point about the transformations of translation and rotation. You can do them in small steps—infinitesimal steps. For example, instead of moving a particle from x to $x+1$, you can move it from x to $x+\delta$. Now I am using δ to denote an infinitesimal. In fact, you can build up the original displacement $x \rightarrow x+1$ by many tiny steps of size δ. The same is true for rotations: You can rotate through an infinitesimal angle δ and, by repeating the process, eventually build up a finite rotation. Transformations like this are called *continuous*: They depend on a continuous parameter (the angle of rotation), and, moreover, you can make the parameter infinitesimal. This will prove to be a good thing, because we can explore all the consequences of continuous symmetries by restricting our attention to the infinitesimal case.

Since finite transformations can be compounded out of infinitesimal ones, in studying symmetries it's enough to consider transformations with very small changes in the coordinates, the so-called *infinitesimal transformations*. So let's consider what happens to Equations (9) when the angle θ is replaced by an infinitesimal angle δ. To first order in δ,

$$\cos\delta = 1$$
$$\sin\delta = \delta.$$

(Recall that for small angles, $\sin\delta = \delta$ and $\cos\delta = 1 - \frac{1}{2}\delta^2$, so the first-order shift in the cosine vanishes and the first-order shift in sine is δ.)

Then the rotation represented by Equations (9) simplifies to

$$x \rightarrow x + y\delta$$
$$y \rightarrow y - x\delta. \tag{10}$$

You can also see that the velocity components change. Just differentiate Equations (10) with respect to time:

$$\dot{x} \rightarrow \dot{x} + \dot{y}\delta$$
$$\dot{y} \rightarrow \dot{y} - \dot{x}\delta. \tag{11}$$

Another way to express the effect of the infinitesimal transformation is to concentrate on the changes in the coordinates and write

$$\delta x = y\delta$$
$$\delta y = -x\delta. \tag{12}$$

Now it's a simple calculus exercise to show that the Lagrangian does not change to first order in δ.

Exercise 4: Show this to be true.

One thing worth noting is that if the potential is not a function of distance from the origin, then the Lagrangian is not invariant with respect to the infinitesimal rotations. This very important point should be checked by examining some explicit examples. A simple example is a potential that depends only on x and not on y.

More General Symmetries

Before we get to the connection between symmetries and conservation laws, let's generalize our notion of symmetry. Suppose the coordinates of an abstract dynamical system are q_i. The general idea of an infinitesimal transformation is that it is a small shift of the coordinates, that may itself depend on the value of the coordinates. The shift is parameterized by an infinitesimal parameter δ, and it has the form

$$\delta q_i = f_i(q)\,\delta. \tag{13}$$

In other words, each coordinate shifts by an amount proportional to δ, but the proportionality factor depends on where you are in configuration space. In the example of Equations (6) the value of f_1 and of f_2 are both 1. In the example of Equations (7) the f-functions are $f_1 = a$ and $f_2 = -b$. But in the more complicated example of the rotations of Equations (12), the f's are not constant:

$$f_x = y$$
$$f_y = -x.$$

If we want to know the change in velocities—in order, for example, to compute the change in the Lagrangian—we need only to differentiate Eq. (13). A little calculus exercise gives

$$\delta \dot{q}_i = \frac{d}{dt}(\delta q_i). \tag{14}$$

For example, from Equations (12),

$$\delta \dot{x} = \dot{y} \delta$$
$$\delta \dot{y} = -\dot{x} \delta. \tag{15}$$

Now we can re-state the meaning of a symmetry for the infinitesimal case. A continuous symmetry is an infinitesimal transformation of the coordinates for which the change in the Lagrangian is zero. It is particularly easy to check whether the Lagrangian is invariant under a continuous symmetry: All you have to do is to check whether the first order variation of the Lagrangian is zero. If it is, then you have a symmetry.

Now let's see what the consequences of a symmetry are.

The Consequences of Symmetry

Let's calculate how much $L(q, \dot{q})$ changes when we do a transformation that shifts q_i by the amount in Eq. (13) and, at the same time, shifts $\dot{q}_i$ by the amount in Eq. (14). All we have to do is compute the change due to varying the $\dot{q}$'s and add it to the change due to varying the q's:

$$\delta L = \sum_i \left(\frac{\partial L}{\partial \dot{q}_i} \delta \dot{q}_i + \frac{\partial L}{\partial q_i} \delta q_i \right). \tag{16}$$

Now we do a bit of magic. Watch it carefully. First, we remember that $\frac{\partial L}{\partial \dot{q}_i}$ is the momentum conjugate to q_i, which we denote p_i.

Thus the first term in Eq. (16) is $\sum_i p_i \delta \dot{q}_i$. Hold on to that while we study the second term, $\dfrac{\partial L}{\partial q_i} \delta q_i$. To evaluate terms of this type, we assume the system is evolving along a trajectory that satisfies the Euler-Lagrange equations

$$\frac{\partial L}{\partial q_i} = \frac{d\, p_i}{d\, t}.$$

Combining the terms, here is what we get for the variation of the Lagrangian:

$$\delta L = \sum_i \left(p_i\, \delta \dot{q}_i + \dot{p}_i\, \delta q_i \right).$$

The final piece of magic is to use the product rule for derivatives:

$$\frac{d\,(F\,G)}{d\,t} = \dot{F}\,G + F\,\dot{G}.$$

Thus we get the result

$$\delta L = \frac{d}{d\,t} \sum_i p_i\, \delta q_i.$$

What does all of this have to do with symmetry and conservation? First of all, by definition, symmetry means that the variation of the Lagrangian is zero. So if Eq. (13) is a symmetry, then $\delta L = 0$ and

$$\frac{d}{d\,t} \sum_i p_i\, \delta q_i = 0.$$

But now we plug in the form of the symmetry operation, Eq. (13), and get

$$\frac{d}{dt} \sum_i p_i \, f_i(q) = 0. \qquad (17)$$

That's it: The conservation law is proved. What Eq. (17) states is that a certain quantity,

$$Q = \sum_i p_i \, f_i(q), \qquad (18)$$

does not change with time. In other words, it is conserved. The argument is both abstract and powerful. It did not depend on the details of the system, but only on the general idea of a symmetry. Now let's turn back to some particular examples in light of the general theory.

Back to Examples

Let's apply Eq. (18) to the examples we studied earlier. In the first example, Eq. (1), the variation of the coordinates in Equations (12) defines both f_1 and f_2 to be exactly 1. Plugging $f_1 = f_2 = 1$ into Eq. (18) gives exactly what we found earlier: $(p_1 + p_2)$ is conserved. But now we can say a far more general thing: *For any system of particles, if the Lagrangian is invariant under simultaneous translation of the positions of all particles, then momentum is conserved.* In fact, this can be applied separately to each spatial component of momentum. If L is invariant under translations along the x axis, then the total x component of momentum is conserved. Thus we see that Newton's third law—action equals reaction—is the consequence of a deep fact about space: *Nothing in the laws of physics changes if everything is simultaneously shifted in space.*

Next let's look at the second example, in which the variations of Equations (12) imply $f_1 = b$, $f_2 = -a$. Again, plugging this result into Eq. (18), we find that the conserved quantity is $b\,p_1 + a\,p_2$.

The last example—rotation—is more interesting. It involves a new conservation law that we haven't met yet. From Eq. (14) we obtain $f_x = y$, $f_y = -x$. This time the conserved quantity involves both coordinates and momenta. It is called l, or *angular momentum*. From Eq. (18) we get

$$l = y\,p_x - x\,p_y.$$

Again, as in the case of translations, there is a deeper thing involved than just the angular momentum of a single particle:

For any system of particles, if the Lagrangian is invariant under simultaneous rotation of the position of all particles, about the origin, then angular momentum is conserved.

Exercise 5: Determine the equation of motion for a simple pendulum of length l swinging through an arc in the x, y plane from an initial angle of θ.

So far, our examples have been very trivial. The Lagrangian formulation is beautiful, elegant, blah blah, but is it really good for solving hard problems? Couldn't you just use $F = m\,a$?

Try it. Here is an example: the double pendulum. A pendulum swings in the x, y plane supported at the origin. The rod of the pendulum is massless, and the bob (weight at the end) is M. To make it simple, let the rod be 1 meter in length and let the bob be 1 kilogram in mass. Next, take another identical pendulum, but suspend it from the bob of the first pendulum, as shown in Figure 3. We can study two cases: with and without a gravitational field.

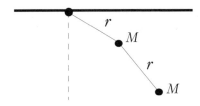

Figure 3: The double pendulum.

Our goal will not be to solve the equations of motion. That we can always do, even if we have to put them on a computer and do it numerically. The goal is to find those equations. It's a tricky problem if you try to do it by $F = M\,a$. Among other things, you have to worry about the forces transmitted through the rod. The Lagrangian method is much easier. There is a more or less mechanical procedure for doing it. The steps are the following:

1. Choose some coordinates that uniquely specify the configuration of the components. You can choose them however you like—just make sure that you have just enough to determine the configuration—and keep them as simple as possible.

    In the double pendulum example, you need two coordinates. I will choose the first one to be the angle of the first pendulum from the vertical. Call it θ. Next, I have a choice. Should I choose the second angle (the angle of the second rod) also to be measured from the vertical, or should I measure it relative to the angle of the first rod? The answer is that it does not matter. One choice may make the equations a little simpler, but either will get you to the answer. I will choose the angle α to be measured relative to the first rod rather than to the vertical.

2.   Work out the total kinetic energy. In this case it is the kinetic energy of the two bobs.

      The easiest way to do this is to refer temporarily to Cartesian coordinates x, y. Let x_1, y_1 refer to the first bob and x_2, y_2 to the second bob. Here are some relations among the angles θ, α and x, y: For bob 1,

$$x_1 = \sin \theta$$
$$y_1 = \cos \theta$$

and for bob 2,

$$x_2 = \sin \theta + \sin (\alpha + \theta)$$
$$y_2 = \cos \theta + \cos (\alpha + \theta).$$

Now, by differentiating with respect to time, you can compute the Cartesian velocity components in terms of the angles and their time derivatives.

      Finally, work out the kinetic energy $\frac{m}{2}\left(\dot{x}^2 + \dot{y}^2\right)$ for each bob and add them. It should take a couple of minutes. Remember that we have chosen the masses and rod lengths to be 1.

      Here is the result: The kinetic energy of the first bob is:

$$T_1 = \frac{\dot{\theta}^2}{2}$$

and the kinetic energy of the second bob is

$$T_2 = \frac{\dot{\theta}^2 + \left(\dot{\theta} + \dot{\alpha}\right)^2}{2} + \dot{\theta}\left(\dot{\theta} + \dot{\alpha}\right) \cos \alpha.$$

If there is no gravitational field, then the kinetic energy is the Lagrangian:

$$L = T_1 + T_2 = \frac{\dot{\theta}^2}{2} + \frac{\dot{\theta}^2 + \left(\dot{\theta} + \dot{\alpha}\right)^2}{2} + \dot{\theta}\left(\dot{\theta} + \dot{\alpha}\right)\cos\alpha.$$

If there is gravity, then we have to calculate the gravitational potential energy. That's easy: For each bob we add its altitude times $m\,g$. This gives a potential energy

$$V(\theta, \alpha) = -g\,[2\cos\theta + \cos(\theta - \alpha)].$$

3. Work out the Euler-Lagrange equations for each degree of freedom.

4. For later purposes, work out the conjugate momenta for each coordinate, $p_i = \dfrac{\partial L}{\partial \dot{q}_i}$.

Exercise 6: Work out the Euler-Lagrange equations for θ and α.

There is more that you may want to do. In particular, you may want to identify the conserved quantities. Energy is usually the first one. The total energy is just $T + V$. But there may be more. Finding symmetries is not always a mechanical procedure; you may have to do some pattern recognition. In the double pendulum case without any gravity, there is another conservation law. It follows from rotation symmetry. Without a gravitational field, if you rotate the whole system about the origin, nothing changes. This implies conservation of angular momentum, but to find the form of the angular momentum, you have to go through the procedure that we derived. That involves knowing the conjugate momenta.

Exercise 7: Work out the form of the angular momentum for the double pendulum, and prove that it is conserved when there is no gravitational field.

Lecture 8: Hamiltonian Mechanics and Time-Translation Invariance

Doc was sitting at the bar drinking his usual—a beer milk shake—and reading the paper, when Lenny and George walked in. "What are you reading about, Doc?"

Doc looked up at Lenny over his glasses. "I see where this guy Einstein says, 'Insanity is doing the same thing over and over and expecting different results.' What do you think about that?"

Lenny thought for a minute. "You mean like every time I eat here I order chili, and then I get a stomachache?"

Doc chuckled, "Yeah, that's the idea. I see you're beginning to understand Einstein."

Time-Translation Symmetry

You may wonder what happened to energy conservation and whether it fits the pattern relating symmetries to conservation laws. Yes, it does, but in a slightly different way than the examples in Lecture 7. In all of those examples the symmetry involved shifting the coordinates q_i. For instance, a translation is a symmetry that simultaneously shifts the Cartesian coordinates of all the particles in a system by the same amount. The symmetry connected with energy conservation involves a shift of *time*.

Imagine an experiment involving a closed far from any perturbing influences. The experiment begins at time t_0

with a certain initial condition, proceeds for a definite period, and results in some outcome. Next, the experiment is repeated in exactly the same way but at a later time. The initial conditions are the same as before, and so is the duration of the experiment; the only difference is the starting time, which is pushed forward to $t_0 + \Delta t$. You might expect that the outcome will be exactly the same, and that the shift Δt would make no difference. Whenever this is true, the system is said to be invariant under *time translation.*

Time-translation invariance does not always apply. For example, we live in an expanding universe. The effect of the expansion on ordinary laboratory experiments is usually negligible, but it's the principle that counts. At some level of accuracy, an experiment that begins later will have a slightly different outcome than one which begins earlier.

Here is a more down-to-earth example. Suppose the system of interest is a charged particle moving in a magnetic field. If the magnetic field is constant then the motion of the particle will be time-translation invariant. But if the current that generates the field is being slowly increased, then the same initial condition for the particle—but starting at different times—will result in a different outcome. The description of the particle will not be time-translation invariant.

How is time-translation symmetry, or the lack of it, reflected in the Lagrangian formulation of mechanics? The answer is simple. In those cases where there is such symmetry, the Lagrangian has no explicit dependence on time. This is a subtle point: The value of the Lagrangian may vary with time, but *only* because the coordinates and velocities vary. Explicit time dependence means that the *form* of the Lagrangian depends on time. For example, take the harmonic oscillator with Lagrangian

$$L = \frac{1}{2}\left(m\dot{x}^2 - kx^2\right).$$

If m and k are time-independent then this Lagrangian is time-translation invariant.

But one can easily imagine that the spring constant k might, for some reason, change with time. For example, if the experiment took place in a changing magnetic field, this could have a subtle effect on the atoms of the spring, which in turn could cause k to vary. In that case, we would have to write

$$L = \frac{1}{2}\left[m\dot{x}^2 - k(t)x^2\right].$$

This is what we mean by an explicit time dependence. More generally, we can write

$$L = L\left(q_i, \dot{q}_i, t\right), \tag{1}$$

where the t dependence is due to the time variation of all the parameters controlling the behavior of the system.

With this idea in hand, we can now give a very succinct mathematical criterion for time-translation symmetry: *A system is time-translation invariant if there is no explicit time dependence in its Lagrangian.*

Energy Conservation

Let's consider how the actual value of the Lagrangian, Eq. (1), changes as a system evolves. There are three sources of time dependence of L. The first and second are due to the time dependence of the coordinates q and the velocities $\dot{q}$. If that were all, we would write

$$\frac{dL}{dt} = \sum_i \left(\frac{\partial L}{\partial q_i} \dot{q}_i + \frac{\partial L}{\partial \dot{q}_i} \ddot{q}_i \right).$$

But if the Lagrangian has explicit time dependence, then there is another term:

$$\frac{dL}{dt} = \sum_i \left(\frac{\partial L}{\partial q_i} \dot{q}_i + \frac{\partial L}{\partial \dot{q}_i} \ddot{q}_i \right) + \frac{\partial L}{\partial t}. \tag{2}$$

Let's examine the various terms in Eq. (2) using the Euler-Lagrange equations of motion. The first type of term, $\dfrac{\partial L}{\partial q_i} \dot{q}_i$, can be written

$$\frac{\partial L}{\partial q_i} \dot{q}_i = \dot{p}_i \dot{q}_i.$$

The second type of term, $\dfrac{\partial L}{\partial \dot{q}_i} \ddot{q}_i$, takes the form

$$\frac{\partial L}{\partial \dot{q}_i} \ddot{q}_i = p_i \ddot{q}_i.$$

If we combine everything, we get

$$\frac{dL}{dt} = \sum_i \left(\dot{p}_i \dot{q}_i + p_i \ddot{q}_i \right) + \frac{\partial L}{\partial t}.$$

The first two terms can be simplified. We use the identity

$$\sum_i \left(\dot{p}_i \dot{q}_i + p_i \ddot{q}_i \right) = \frac{d}{dt} \sum_i \left(p_i \dot{q}_i \right)$$

to get

$$\frac{dL}{dt} = \frac{d}{dt} \sum_i \left(p_i \dot{q}_i \right) + \frac{\partial L}{\partial t}. \tag{3}$$

Notice that even if there is no explicit time dependence in L, the Lagrangian will nevertheless depend on time through the first term $\sum_i \frac{d}{dt} \left(p_i \dot{q}_i \right)$. The upshot is that there is no such thing as conservation of the Lagrangian.

Inspection of Eq. (3) reveals something interesting. If we define a new quantity H by

$$\sum_i \left(p_i \dot{q}_i \right) - L = H \tag{4}$$

then Eq. (3) has a very simple form:

$$\frac{dH}{dt} = -\frac{\partial L}{\partial t}. \tag{5}$$

The steps leading to Eq. (5) may seem a bit complicated, but the result is very simple. The new quantity H varies with time *only* if the Lagrangian has an explicit time dependence. An even more interesting way to say it is *if a system is time-translation invariant, then the quantity H is conserved.*

The quantity H is called the *Hamiltonian*, and, as you might expect, it is important because (among other reasons) it is the energy of a system. But it is more than important; it is the central element in an entirely new formulation of mechanics called the *Hamiltonian formulation*. But for now, let's consider its meaning by returning to a simple example, the motion of a particle in a potential. The Lagrangian is

$$L = \frac{m}{2}\dot{x}^2 - V(x), \tag{6}$$

and the canonical momentum is just the usual momentum

$$p = m\dot{x}. \tag{7}$$

Let's plug Eq. (6) and Eq. (7) into Eq. (4), the definition of H:

$$H = \left(m\dot{x}\right)\dot{x} - \frac{m}{2}\dot{x}^2 + V(x)$$

$$= m\dot{x}^2 - \frac{m}{2}\dot{x}^2 + V(x)$$

$$= \frac{m}{2}\dot{x}^2 + V(x).$$

Notice what happens: Two terms proportional to $m\dot{x}^2$ combine to give the usual kinetic energy, and the potential term becomes $+V(x)$. In other words, H just becomes the usual total energy, kinetic *plus* potential.

This is the general pattern that you can check for any number of particles. If the Lagrangian is kinetic energy minus potential energy, then

$$H = p\dot{q} - T + V.$$
$$= T + V.$$

There are systems for which the Lagrangian has a more intricate form than just $T - V$. For some of those cases, it is not possible to identify a clear separation into kinetic and potential energy. Nonetheless, the rule for constructing the Hamiltonian is the same. The general definition of energy for these systems is

Energy equals Hamiltonian.

Moreover, if there is no explicit time dependence in the Lagrangian, then the energy H is conserved.

If, however, the Lagrangian is explicitly time-dependent, then Eq. (5) implies that the Hamiltonian is not conserved. What happened to the energy in that case? To understand what is going on, let's consider an example. Suppose that a charged particle, with unit electric charge, is moving between the plates of a capacitor. The capacitor has a uniform electric field ϵ due to the charges on the plates. (The reason we are using ϵ for electric field, instead of the more conventional E, is to avoid confusing it with energy.) You don't have to know anything about electricity. All you need to know is that the capacitor creates a potential energy equal to ϵx. The Lagrangian is

$$L = \frac{m}{2}\dot{x}^2 - \epsilon x.$$

As long as the field is constant, the energy is conserved. But suppose the capacitor is being charged up so that ϵ is also ramping up. Then the Lagrangian has an explicit time dependence:

$$L = \frac{m}{2}\dot{x}^2 - \epsilon(t)\,x.$$

Now the energy of the particle is not conserved. Depending on the momentary location x of the particle, the energy varies according to

$$\frac{dH}{dt} = \frac{d\epsilon}{dt}x.$$

Where did that energy come from? The answer is that it came from the battery that was charging the capacitor. I won't go into details, but the point is that when we defined the system to consist of just the particle, we narrowed our focus to just a part of a bigger system that includes the capacitor and the battery. These additional items are also made of particles and therefore have energy.

Consider the entire experiment, including the battery, capacitor, and particle. The experiment begins with an uncharged capacitor and a particle at rest, somewhere between the plates. At some moment we close a circuit, and current flows into the capacitor. The particle experiences a time-dependent field, and, at the end of the experiment, the capacitor is charged and the particle is moving.

What if we did the entire experiment an hour later? The outcome, of course, would be the same. In other words, the entire closed system is time-translation invariant, so the entire energy of all items is conserved. If we treated the entire collection as a single system, it would be time-translation invariant, and the total energy would be conserved.

Nevertheless, it is often useful to divide a system into parts and to focus on one part. In that case, the energy of part of the system will not be conserved if the other parts are varying with time.

Phase Space and Hamilton's Equations

The Hamiltonian is important because (among other reasons) it is the energy. But its significance is far deeper: It is the basis for a complete revamping of classical mechanics, and it is even more important in quantum mechanics.

In the Lagrangian—or action—formulation of mechanics, the focus is on the trajectory of a system through the configuration space. The trajectory is described in terms of the coordinates $q(t)$. The equations are second-order differential equations, so it is not enough to know the initial coordinates; we also have to know the initial velocities.

In the Hamiltonian formulation, the focus is on phase space. Phase space is the space of both the coordinates q_i and the conjugate momenta p_i. In fact, the q's and p's are treated on the same footing, the motion of a system being described by a trajectory through the phase space. Mathematically, the description is through a set of functions $q_i(t)$, $p_i(t)$. Notice that the number of dimensions of phase space is twice that of configuration space.

What do we gain by doubling the number of dimensions? The answer is that the equations of motion become first-order differential equations. In less technical terms, this means that the future is laid out if we know only the initial point in phase space.

The first step in constructing the Hamiltonian formulation is to get rid of the $\dot{q}$'s and replace them with the p's. The goal is to express the Hamiltonian as a function of q's and p's. For particles in ordinary Cartesian coordinates, the momenta and velocities are almost the same thing, differing only by a factor of the mass. As usual, the particle on a line is a good illustration.

We start with the two equations

$$p = m\dot{x}$$

$$H = \frac{m\dot{x}^2}{2} + V(x). \tag{8}$$

When we replace the velocity with p/m the Hamiltonian becomes a function of p and x:

$$H = \frac{p^2}{2\,m} + V(x).$$

One last point before we write the equations of motion in Hamiltonian form: The partial derivative of H with respect to x is just $\frac{d\,V}{d\,x}$, or minus the force. Thus the equation of motion $(F = m\,a)$ takes the form

$$\dot{p} = -\frac{\partial\,H}{\partial\,x}.\qquad(9)$$

We noted earlier that in the Hamiltonian formulation, the coordinates and momenta are on the same footing. From that you might guess that there is another equation similar to Eq. (9), with p and x interchanged. That is almost true, but not quite. The correct equation is

$$\dot{x} = \frac{\partial\,H}{\partial\,p},\qquad(10)$$

with a plus sign instead of a minus sign.

To see why Eq. (10) is true, just differentiate the expression for H with respect to p. From the second of Equations (8) we get

$$\frac{\partial H}{\partial p} = \frac{p}{m},$$

which from the first equation is just $\dot{x}$.

So now we see a very simple symmetric packaging of the equations. We have two equations of motion instead of one, but each is of first-order:

$$\dot{p} = -\frac{\partial H}{\partial x}$$

$$\dot{x} = \frac{\partial H}{\partial p}. \tag{11}$$

These are Hamilton's equations for a particle on a line. Soon we will derive the general form for any system, but for now I will tell you what it is. We start with a Hamiltonian that is a function of all the q's and p's:

$$H = H(q_i, p_i).$$

We can use this to generalize Equations (11),

$$\dot{p}_i = -\frac{\partial H}{\partial q_i}$$

$$\dot{q}_i = \frac{\partial H}{\partial p_i}. \tag{12}$$

So we see that for each direction in phase space, there is a single first-order equation.

Let's stop to consider how these equations are related to the very first chapter of this book, in which we described how deterministic laws of physics predict the future. What Equations (12) say is this:

If at any time you know the exact values of all the coordinates and momenta, and you know the form of the Hamiltonian, Hamilton's equations will tell you the corresponding quantities an infinitesimal time later. By a process of successive updating, you can determine a trajectory through phase space.

The Harmonic Oscillator Hamiltonian

The harmonic oscillator is by far the most important simple system in physics. It describes all sorts of oscillations in which some degree of freedom is displaced and then oscillates about an equilibrium position. To see why it is so important, let's suppose a degree of freedom q has a potential energy $V(q)$ that has a minimum. The minimum describes a stable equilibrium, and when the degree of freedom is displaced, it will tend to return to the equilibrium position. Without any real loss of generality, we can locate the minimum at $q = 0$. The generic function that has a minimum at this point can be approximated by the quadratic function

$$V(q) = V(0) + c\, q^2. \tag{13}$$

where $V(0)$ and c are constants. The reason why there is no linear term proportional to q is that the derivative $\dfrac{dV}{dq}$ must be zero at the minimum. We can also drop the term $V(0)$ since adding a constant to the potential energy has no effect.

The form of Eq. (13) is not very general; V could contain terms of all orders—for example, q^3 or q^4. But as long as the system deviates from $q = 0$ by only a small amount, these higher-order terms will be negligible compared to the quadratic term. This reasoning applies to all sorts of systems: springs, pendulums, oscillating sound waves, electromagnetic waves, and on and on.

I will write the Lagrangian in what may seem like a special form involving a single constant called ω:

$$L = \frac{1}{2\,\omega}\,\dot{q}^2 - \frac{\omega}{2}\,q^2. \tag{14}$$

Exercise 1: Start with the Lagrangian $\frac{m\,\dot{x}^2}{2} - \frac{k}{2}\,x^2$ **and show that if you make the change in variables** $q = (k\,m)^{1/4}\,x$, **the Lagrangian has the form of Eq. (14). What is the connection among** k, m, **and** ω?

Exercise 2: Starting with Eq. (14), calculate the Hamiltonian in terms of p **and** q.

The Hamiltonian corresponding to Eq. (14) is very simple:

$$H = \frac{\omega}{2}\left(p^2 + q^2\right). \tag{15}$$

It was in order to get H into such a simple form that we changed variables from x to q in Exercise 1.

One of the hallmarks of the Hamiltonian formulation is how symmetric it is between the q's and p's. In the case of the harmonic oscillator, it is almost completely symmetric. The only asymmetry is a minus sign in the first of Equations (12). For a single degree of freedom, Hamilton's equations are Equations (11). If we plug our Hamiltonian, Eq. (15), into Equations (12), we get,

$$\dot{p}_i = -\omega\, q$$
$$\dot{q}_i = \omega\, p. \tag{16}$$

How do these two equations compare with Lagrange's equations that we would derive from Eq. (14)? First of all, there is only one Lagrangian equation:

$$\ddot{q} = -\omega^2\, q. \tag{17}$$

Second, this equation is second-order, meaning that it involves second time-derivatives. By contrast, the Hamiltonian equations are each first-order. This somehow means that two first-order equations are equivalent to one second-order equation. We can see this by differentiating the second equation in Equations (16) with respect to time,

$$\ddot{q} = \omega\, \dot{p},$$

and then using the first equation in Equations (16). This enables us to replace $\dot{p}$ with $-\omega\, q$, which gives us Eq. (17): The Euler-Lagrange equation of motion.

Is one formulation better than the other? Did Lagrange have the final word or did Hamilton? You can decide for yourself, but wait a while before you do. We still have a couple of courses on relativity and quantum mechanics before the real meanings of the Lagrangian and Hamiltonian become completely clear.

Let's return to Equations (16). We usually "think" in configuration space. The harmonic oscillator is a system that moves back and forth along a single axis. But it is also an excellent starting point for getting used to "thinking" in phase space. Phase space (for the oscillator) is two-dimensional. It is easy to see that the trajectories of the oscillator in phase space are concentric circles about the origin. The argument is very simple. Go back to the expression for the Hamiltonian, Eq. (15). The Hamiltonian, being the energy, is conserved. It follows that $q^2 + p^2$ is constant with time. In other words, the distance from the origin of phase space is constant, and the phase point moves on a circle of fixed radius. In fact Eq. (16) is the equation for a point moving with constant angular velocity ω about the origin. Especially interesting is the fact that the angular velocity in phase space is the same for all orbits, independent of the energy of the oscillator. As the phase point circles the origin, you can project the motion onto the horizontal q axis, as shown in Figure 1. It moves back and forth in an oscillatory motion, exactly as expected. However, the two-dimensional circular motion through phase space is a more comprehensive description of the motion. By projecting onto the vertical p axis, we see that the momentum also oscillates.

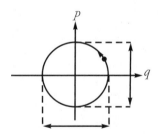

Figure 1: The harmonic oscillator in phase space.

The harmonic oscillator is especially simple. In general, the motion of a system through phase space is more complicated and less symmetric. But the fact that the phase point stays on a contour of constant energy is universal. Later we will discover more general properties of motion in phase space.

Derivation of Hamilton's Equations

Let's complete a piece of business that we left unfinished: the general derivation of Hamilton's equations. The Lagrangian is some general function of the coordinates and velocities,

$$L = L\big(\{q\}, \{\dot{q}\}\big),$$

and the Hamiltonian is

$$H = \sum_i \big(p_i\, \dot{q}_i\big) - L.$$

The change in the Hamiltonian is

$$\delta H = \sum_i \big(p_i\, \delta \dot{q}_i + \dot{q}_i\, \delta p_i\big) - \delta L$$

$$= \sum_i \left(p_i\, \delta \dot{q}_i + \dot{q}_i\, \delta p_i - \frac{\partial L}{\partial q_i}\, \delta q_i - \frac{\partial L}{\partial \dot{q}_i}\, \delta \dot{q}_i \right).$$

Now if we use the definition of p_i, namely $p_i = \dfrac{\partial L}{\partial \dot{q}_i}$, we see that

the first and last terms exactly cancel, leaving

$$\delta H = \sum_i \left(\dot{q}_i\, \delta p_i - \frac{\partial L}{\partial q_i}\, \delta q_i \right).$$

Let's compare this with the general rule for a small change in a function of several variables:

$$\delta H(\{q\}, \{p\}) = \sum_i \left(\frac{\partial H}{\partial p_i}\, \delta p_i + \frac{\partial H}{\partial q_i}\, \delta q_i \right).$$

By matching the terms proportional to δq_i and δp_i, we arrive at

$$\frac{\partial H}{\partial p_i} = \dot{q}_i$$

$$\frac{\partial H}{\partial q_i} = -\frac{\partial L}{\partial \dot{q}_i}. \tag{18}$$

There is only one last step, and that is to write Lagrange's equations in the form

$$\frac{\partial L}{\partial \dot{q}_i} = p_i.$$

Inserting this in the second of Equations (18) we get Hamilton's equations,

$$\frac{\partial H}{\partial p_i} = \dot{q}_i$$

$$\frac{\partial H}{\partial q_i} = -\dot{p}_i. \tag{19}$$

Lecture 9: The Phase Space Fluid and the Gibbs-Liouville Theorem

Lenny loved watching the river, especially watching little bits of floating debris making their way downstream. He tried to guess how they would move between the rocks or get caught in eddys. But the river as a whole—the large-scale current, the volume of water, the shear, and the divergence and convergence of the flow were beyond him.

The Phase Space Fluid

Focusing on a particular initial condition and following it along its specific trajectory through phase space are very natural things to do in classical mechanics. But there is also a bigger picture that emphasizes the entire collection of trajectories. The bigger picture involves visualizing all possible starting points and all possible trajectories. Instead of putting your pencil down at a point in phase space and then following a single trajectory, try to do something more ambitious. Imagine you had an infinite number of pencils and used them to fill phase space uniformly with dots (by uniformly, I mean that the density of dots in the q, p space is everywhere the same). Think of the dots as particles that make up a fictitious phase-space-filling fluid.

Then let each dot move according to the Hamiltonian equations of motion,

$$\dot{q}_i = \frac{\partial H}{\partial p_i}$$

$$\dot{p}_i = -\frac{\partial H}{\partial q_i}, \tag{1}$$

so that the fluid endlessly flows through the phase space.

The harmonic oscillator is a good example to start with. In Lecture 8 we saw that each dot moves in a circular orbit with uniform angular velocity. (Remember, we are talking about phase space, not coordinate space. In coordinate space, the oscillator moves back and forth in one dimension.) The whole fluid moves in a rigid motion, uniformly circulating around the origin of phase space.

Now let's return to the general case. If the number of coordinates is N, then the phase space, and the fluid, are $2N$-dimensional. The fluid flows, but in a very particular way. There are features of the flow that are quite special. One of these special features is that if a point starts with a given value of energy—a given value of $H(q, p)$—then it remains with that value of energy. The surfaces of fixed energy (for example, energy E) are defined by the equation

$$H(q, p) = E. \tag{2}$$

For each value of E we have a single equation for $2N$ phase-space variables, thus defining a surface of dimension $2N - 1$. In other words, there is a surface for each value of E; as you scan over values of E, those surfaces fill up the phase space. You can think of the phase space, along with the surfaces defined in Eq. (2) as a contour map (see Figure 1), but, instead of representing altitude, the contours denote the value of the energy. If a point of the fluid is on a particular surface, it stays on that surface forever. That's energy conservation.

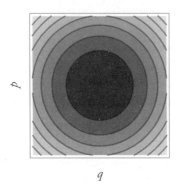

Figure 1: Contour plot of energy surfaces of a harmonic oscillator in phase space.

For the harmonic oscillator, the phase space is two-dimensional and the energy surfaces are circles:

$$\frac{\omega}{2}\left(q^2 + p^2\right) = E. \tag{3}$$

For a general mechanical system, the energy surfaces are far too complicated to visualize, but the principle is the same: *The energy surfaces fill the phase space like layers and the flow moves so that the points stay on the surface that they begin on.*

A Quick Reminder

We want to stop here and remind you of the very first lecture, where we discussed coins, dice, and the simplest idea of a law of motion. We described those laws by a set of arrows connecting dots that represented the states of the system. We also explained that there are allowable laws and unallowable laws, the allowable laws being reversible. What is it that characterizes an allowable law? The answer is that every point must have exactly one incoming arrow and one outgoing arrow. If at any point the number of incoming arrows exceeds the number of outgoing arrows (such a situation is called a *convergence*), then the law is irreversible. The same is true if the number of outgoing arrows exceeds the number of incoming arrows (such a situation is called a *divergence*). Either a convergence or divergence of the arrows violates reversibility and is forbidden. So far we have not returned to that line of reasoning. Now is the time.

Flow and Divergence

Let's consider some simple examples of fluid flow in ordinary space. Forget about phase space for the moment, and just consider an ordinary fluid moving through regular three-dimensional space labeled by axes x, y, z. The flow can be described by a *velocity field*. The velocity field $\vec{v}(x, y, z)$ is defined by going to each point of space and specifying the velocity vector at that point (see Figure 2).

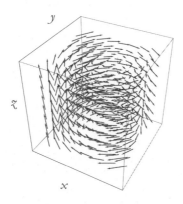

Figure 2: Velocity field.

Or we may describe the velocity field to be the components of the velocity: $v_x(x, y, z)$, $v_y(x, y, z)$, $v_z(x, y, z)$. The velocity at a point might also depend on time, but let's suppose that it doesn't. In that case the flow is called *stationary*.

Now let's suppose the fluid is incompressible. This means that a given amount of fluid always occupies the same volume. It also means that the density of the fluid—the number of molecules per unit volume—is uniform and stays that way forever. By the way, the term incompressible also means in-decompressible. In other words, the fluid cannot be stretched out, or decompressed. Consider a small cubic box defined by

$$x_0 < x < x_0 + dx$$
$$y_0 < y < y_0 + dy$$
$$z_0 < z < z_0 + dz.$$

Incompressibility implies that the number of fluid points in every such box is constant. It also means that the net flow of fluid into the box (per unit time) must be zero. (As many points flow in as flow out.) Consider the number of molecules per unit time coming into the box across the face $x = x_0$. It will be proportional to the flow velocity across that face, $v_x(x_0)$.

If v_x were the same at x_0 and at $x_0 + dx$, then the flow into the box at $x = x_0$ would be the same as the flow out of the box at $x = x_0 + dx$. However, if v_x varies across the box, then the two flows will not balance. Then the net flow into the box across the two faces will be proportional to

$$-\frac{\partial v_x}{\partial x} dx\, dy\, dz.$$

Exactly the same reasoning applies to the faces at y_0 and $y_0 + dy$, and also at z_0 and $z_0 + dz$. In fact, if you add it all up, the net flow of molecules into the box (inflow minus outflow) is given by

$$-\left(\frac{\partial v_x}{\partial x} + \frac{\partial v_y}{\partial y} + \frac{\partial v_z}{\partial z} \right) d x d y d z.$$

The combination of derivatives in the parentheses has a name: It is the *divergence* of the vector field $\vec{v}(t)$ and is denoted by

$$\nabla \cdot \vec{v} = \left(\frac{\partial v_x}{\partial x} + \frac{\partial v_y}{\partial y} + \frac{\partial v_z}{\partial z} \right). \tag{4}$$

The divergence is aptly named; it represents a spreading out of the molecules, or an increase in the volume occupied by the molecules. If the fluid is incompressible, then the volume must not change, and this implies that the divergence must be zero.

One way to think about incompressibility is to imagine that each of the molecules, or points, of the fluid occupies a volume that cannot be compromised. They cannot be squeezed into a smaller volume, nor can they disappear or appear from nowhere. With a little bit of thought, you can see how similar incompressibility is to reversibility. In the examples that we examined in Lecture 1, the arrows also defined a kind of flow. And in a sense the flow was incompressible, at least if it was reversible. The obvious question that this raises is whether the flow through phase space is incompressible. The answer is yes, if the system satisfies Hamilton's equations. And the theorem that expresses the incompressibility is called Liouville's theorem.

Liouville's Theorem

Let's go back to the fluid flow in phase space and consider the components of the velocity of the fluid at every point of the phase space. Needless to say, the phase-space fluid is not three-dimensional with coordinates x, y, z. Instead it is a $2N$-dimensional fluid with coordinates p_i, q_i. Therefore, there are $2N$ components of the velocity field, one for each q and one for each p. Let's call them v_{q_i} and v_{p_i}.

The concept of a divergence in Eq. (4) is easily generalized to any number of dimensions. In three dimensions it is the sum of the derivatives of the velocity components in the respective directions. It's exactly the same in any number of dimensions. In the case of phase space, the divergence of a flow is the sum of $2N$ terms:

$$\nabla \cdot \vec{v} = \sum_i \left(\frac{\partial v_{q_i}}{\partial q_i} + \frac{\partial v_{p_i}}{\partial p_i} \right). \qquad (5)$$

If the fluid is incompressible, then the expression in Eq. (5) must be zero. To find out, we need to know the components of the velocity field—that being nothing but the velocity of a particle of the phase space fluid.

The flow vector of a fluid at a given point is identified with the velocity of a fictitious particle at that point. In other words,

$$v_{q_i} = \dot{q}_i$$
$$v_{p_i} = \dot{p}_i.$$

Moreover, $\dot{q}_i$ and $\dot{p}_i$ are exactly the quantities that Hamilton's equations, Equations (1), give:

$$v_{q_i} = \frac{\partial H}{\partial p_i}$$

$$v_{p_i} = -\frac{\partial H}{\partial q_i}. \tag{6}$$

All we have to do is plug Equations (6) into Eq. (5) and see what we get:

$$\nabla \cdot \vec{v} = \sum_i \left(\frac{\partial}{\partial q_i} \frac{\partial H}{\partial p_i} - \frac{\partial}{\partial p_i} \frac{\partial H}{\partial q_i} \right). \tag{7}$$

Recalling that a second derivative like $\dfrac{\partial}{\partial q_i} \dfrac{\partial H}{\partial p_i}$ does not depend on the order of differentiation, we see that the terms in Eq. (7) exactly cancel in pairs:

$$\nabla \cdot \vec{v} = 0.$$

Thus the phase space fluid is incompressible. In classical mechanics, the incompressibility of the phase space fluid is called *Liouville's theorem*, even though it had very little to do with the French mathematician Joseph Liouville. The great American physicist Josiah Willard Gibbs first published the theorem in 1903, and it is also known as the Gibbs-Liouville theorem.

We defined the incompressibity of a fluid by requiring that the total amount of fluid that enters every small box be zero. There is another definition that is exactly equivalent. Imagine a volume of fluid at a given time. The volume of fluid may have any shape—a sphere, a cube, a blob, or whatever. Now follow all the points in that volume as they move. After a time the fluid blob will be at a different place with a different shape. But if the fluid is incompressible, the volume of the blob will remain what it was at the beginning. Thus we can rephrase Liouville's theorem: *The volume occupied by a blob of phase space fluid is conserved with time.*

Let's take the example of the harmonic oscillator in which the fluid moves around the origin in circles. It's obvious that a blob maintains its volume since all it does is rigidly rotate. In fact, the shape of the blob stays the same. But this latter fact is special to the harmonic oscillator. Let's take another example. Suppose the Hamiltonian is given by

$$H = p\,q.$$

You probably don't recognize this Hamiltonian, but it is completely legitimate. Let's work out its equations of motion:

$$\dot{q} = q$$
$$\dot{p} = -p.$$

What these equations say is that q increases exponentially with time, and p decreases exponentially at the same rate. In other words, the flow compresses the fluid along the p axis, while expanding it by the same amount along the q axis. Every blob gets stretched along q and squeezed along p. Obviously, the blob undergoes an extreme distortion of its shape—but its phase space volume does not change.

Liouville's theorem is the closest analogy that we can imagine to the kind of irreversibility we discussed in Lecture 1. In quantum mechanics, Liouville's theorem is replaced by a quantum version called *unitarity*. Unitarity is even more like the discussion in Lecture 1—but that's for the next installment of *The Theoretical Minimum.*

Poisson Brackets

What were the nineteenth-century French mathematicians thinking when they invented these extremely beautiful—and extremely formal—mathematical ways of thinking about mechanics? (Hamilton himself was an exception—he was Irish.) How did they get the action principle, Lagrange's equations, Hamiltonians, Liouville's theorem? Were they solving physics problems? Were they just playing with the equations to see how pretty they could make them? Or were they devising principles by which to characterize new laws of physics? I think the answer is a bit of each, and they were incredibly successful in all these things. But the really astonishing degree of success did not become apparent until the twentieth century when quantum mechanics was discovered. It almost seems as if the earlier generation of mathematicians were clairvoyant in the way they invented exact parallels of the later quantum concepts.

And we are not finished. There is one more formulation of mechanics that seems to have been very prescient. We owe it to the French mathematician Poisson, whose name means "fish" in French. To motivate the concept of a Poisson bracket, let's consider some function of q_i and p_i. Examples include the kinetic energy of a system that depends on the p's, the potential energy that depends on the q's, or the angular momentum that depends on products of p's and q's. There are, of course, all sorts of other quantities that we might be interested in. Without specifying the particular function, let's just call it $F(q, p)$.

We can think of $F(q, p)$ in two ways. First of all, it is a function of position in the phase space. But if we follow any point as it moves through the phase space—that is, any actual trajectory of the system—there will be a value of F that varies along the trajectory. In other words, the motion of the system along a particular trajectory turns F into a function of time. Let's compute how F varies for a given point as it moves, by computing the time derivative of F:

$$\dot{F} = \sum_i \left(\frac{\partial F}{\partial q_i} \dot{q}_i + \frac{\partial F}{\partial p_i} \dot{p}_i \right).$$

By now the routine should be obvious—we use Hamilton's equations for the time derivatives of q and p:

$$\dot{F} = \sum_i \left(\frac{\partial F}{\partial q_i} \frac{\partial H}{\partial p_i} - \frac{\partial F}{\partial p_i} \frac{\partial H}{\partial q_i} \right). \tag{8}$$

I don't know exactly what Poisson was doing when he invented his bracket, but I suspect he just got tired of writing the right hand side of Eq. (8) and decided to abbreviate it with a new symbol. Take any two functions of phase space, $G(q, p)$ and $F(q, p)$. Don't worry about their physical meaning or whether one of them is the Hamiltonian. The Poisson bracket of F and G is defined as

$$\{F, G\} = \sum_i \left(\frac{\partial F}{\partial q_i} \frac{\partial G}{\partial p_i} - \frac{\partial F}{\partial p_i} \frac{\partial G}{\partial q_i} \right). \tag{9}$$

Poisson could now save himself the trouble of writing Eq. (8). Instead, he could write

$$\dot{F} = \{F, H\}. \tag{10}$$

The amazing thing about Eq. (10) is that it summarizes so much. The time derivative of anything is given by the Poisson bracket of that thing with the Hamiltonian. It even contains Hamilton's equations themselves. To see that, let $F(q, p)$ just be one of the q's:

$$\dot{q}_k = \{q_k, H\}.$$

Now, if you work out the Poisson bracket of q_i and H, you will discover that it has only one term—namely, the one where you differentiate q_k with respect to itself. Since $\dfrac{\partial q_k}{\partial q_k} = 1$, we find that the Poisson bracket $\{q_k, H\}$ is just equal to $\dfrac{\partial H}{\partial p}$, and we recover the first of Hamilton's equations. The second equation is easily seen to be equivalent to

$$\dot{p}_k = \{p_k, H\}.$$

Notice that in this formulation the two equations have the same sign, the sign difference being buried in the definition of the Poisson bracket.

The French obsession with elegance really paid off. The Poisson bracket turned into one of the most basic quantities of quantum mechanics: the commutator.

Lecture 10: Poisson Brackets, Angular Momentum, and Symmetries

Lenny asked, "Hey George, can we hang fish on a Poisson Bracket?"

George smiled. "Only if they're theoretical."

An Axiomatic Formulation of Mechanics

Let's abstract a set of rules that enable one to manipulate Poisson Brackets (from now on I'll use the abbreviation PB) without all the effort of explicitly calculating them. You can check (consider it homework) that the rules really do follow from the definition of PB's. Let A, B, and C be functions of the p's and q's. In the last lecture, I defined the PB:

$$\{A, C\} = \sum_i \left(\frac{\partial A}{\partial q_i} \frac{\partial C}{\partial p_i} - \frac{\partial A}{\partial p_i} \frac{\partial C}{\partial q_i} \right). \tag{1}$$

- The first property is *antisymmetry*: If you interchange the two functions in the PB it changes sign:

$$\{A, C\} = -\{C, A\}. \tag{2}$$

In particular, that means that the PB of a function with itself is zero:

$$\{A, A\} = 0. \tag{3}$$

- Next is *linearity* in either entry. Linearity entails two properties. First, if you multiply A (but not C) by a constant k, the PB gets multiplied by the same constant:

$$\{k\,A, C\} = k\,\{A, C\}. \tag{4}$$

Second if you add $A + B$ and take the PB with C, the result is additive:

$$\{(A + B), C\} = \{A, C\} + \{B, C\}. \tag{5}$$

Eq.s (2) and (3) define the linearity property of PB's.

- Next we consider what happens when we multiply A and B and then take the PB with C. To figure it out, all you need to do is go back to the definition of the PB and apply the rule for differentiating a product. For example,

$$\frac{\partial(A\,B)}{\partial q} = A\,\frac{\partial B}{\partial q} + B\,\frac{\partial A}{\partial q}.$$

The same thing is true for derivatives with respect to p. Here is the rule:

$$\{(A\,B), C\} = B\,\{A, C\} + A\,\{B, C\}. \tag{6}$$

- Finally, there are some specific PB's that you need in order to get started. Begin by noting that any q or any p is a function of the p's and q's. Since every PB involves derivatives with respect to both p's and q's, the PB of any q with any other q is zero. The same is true for the PB of two p's:

$$\begin{aligned}\{q_i, q_j\} &= 0 \\ \{p_i, p_j\} &= 0. \end{aligned} \tag{7}$$

But a PB of a q with a p is not zero. The rule is that $\{q_i, p_j\}$ is one if $i = j$ and zero otherwise. Using the Kronecker symbol,

$$\{q_i, p_j\} = \delta_{ij}. \tag{8}$$

Now we have everything we need to calculate any PB. We can forget the definition and think of Eq.s (2, 3, 4, 5, 6, 7, and 8) as a set of axioms for a formal mathematical system.

Suppose we want to compute

$$\{q^n, p\}, \tag{9}$$

where for simplicity I have assumed a system with just one q and one p. I will tell you the answer and then prove it. The answer is

$$\{q^n, p\} = n\, q^{(n-1)}. \tag{10}$$

The way to prove this kind of formula is to use *mathematical induction*. That takes two steps. The first step is to assume the answer for n (assume the induction hypothesis, Eq. (10)) and show that it follows for $n + 1$. The second step is to explicitly show that the induction hypothesis holds for $n = 1$.

Thus, replacing n with $n + 1$, we can write Eq. (9) using Eq. (6):

$$\begin{aligned}
\{q^{(n+1)}, p\} &= \{q \cdot q^n, p\} \\
&= q\{q^n, p\} + q^n\{q, p\}.
\end{aligned}$$

Next use Eq. (8), which in this case is just $\{q, p\} = 1$:

$$\begin{aligned}
\{q^{(n+1)}, p\} &= \{q \cdot q^n, p\} \\
&= q\{q^n, p\} + q^n.
\end{aligned}$$

We now use the induction hypothesis—Eq. (10)—and get

$$\begin{aligned}
\{q^{(n+1)}, p\} &= \{q \cdot q^n, p\} \\
&= q\, n\, q^{(n-1)} + q^n \tag{11} \\
&= (n + 1)\, q^n.
\end{aligned}$$

Equation (11) is exactly the induction hypothesis for $n + 1$. Therefore, all we need to do is show that Eq. (10) holds for

$n = 1$. But all it says is that $\{q, p\} = 1$, which is of course true. Thus Eq. (10) is true.

We can write this example in another way that has far-reaching consequences. Notice that $n q^{(n-1)}$ is nothing but the derivative of q^n. Thus, for this case,

$$\{q^n, p\} = \frac{d(q^n)}{d q}. \tag{12}$$

Now take any polynomial (even an infinite power series) of q. By applying Eq. (12) to each term in the polynomial and using linearity to combine the results, we can prove

$$\{F(q), p\} = \frac{d F(q)}{d q}. \tag{13}$$

Since any smooth function can be arbitrarily well approximated by a polynomial, this enables us to prove Eq. (13) for any function of q. In fact, it even goes further. For any function of q and p, it is easy to prove that

$$\{F(q, p), p_i\} = \frac{\partial F(q, p)}{\partial q_i}. \tag{14}$$

Exercise 1: Prove Eq. (14).

Thus we have discovered a new fact about Poisson Brackets: *Taking the PB of any function with p_i has the effect of differentiating the function with respect to q_i.* We could have proved that directly from the definition of the PB, but I wanted to show you that it follows from the formal axioms.

What about taking the Poisson bracket of $F(q, p)$ with q_i? You may be able to guess the answer from the symmetric way in which the p's and q's enter all the rules. By now you may even guess the sign of the answer:

$$\{F(q, p), q_i\} = -\frac{\partial F(q, p)}{\partial p_i}. \tag{15}$$

Exercise 2: Hamilton's equations can be written in the form $\dot{q} = \{q, H\}$ and $\dot{p} = \{p, H\}$. Assume that the Hamiltonian has the form $H = \frac{1}{2m} p^2 + V(q)$. Using only the PB axioms, prove Newton's equations of motion.

Angular Momentum

In Lecture 7, I explained the relationship between rotation symmetry and the conservation of angular momentum. Just to remind you, I will briefly review it for the case of a single particle moving in the x, y plane. We wrote the formula for an infinitesimal rotation in the form

$$\begin{aligned} \delta x &= \epsilon f_x = -\epsilon y \\ \delta y &= \epsilon f_y = \epsilon x. \end{aligned} \tag{16}$$

Then, assuming that the Lagrangian is invariant, we derived a conserved quantity

$$Q = p_x f_x + p_y f_y;$$

with a change of sign, we call it the angular momentum L,

$$L = x\, p_y - y\, p_x. \tag{17}$$

Now I want to go to three-dimensional space, where angular momentum has the status of a vector. Equation (16) is still true, but it takes on a new meaning: It becomes the rule for rotating a system about the z axis. In fact, we can fill it out with a third equation that expresses the fact that z is unchanged by a rotation about the z axis:

$$
\begin{aligned}
\delta x &= \epsilon\, f_x = -\epsilon\, y \\
\delta y &= \epsilon\, f_y = \epsilon\, x \\
\delta z &= 0.
\end{aligned}
\tag{18}
$$

Equation (17) is also unchanged, except that we interpret the left-hand side as the z component of the angular momentum. The other two components of angular momentum are also easily computed, or you can guess them just by cycling the equation $x \to y, y \to z, z \to x$:

$$
\begin{aligned}
L_z &= x\, p_y - y\, p_x \\
L_x &= y\, p_z - z\, p_y \\
L_y &= z\, p_x - x\, p_z.
\end{aligned}
$$

As you might expect, each component of the vector $\vec{L}$ is conserved if the system is rotationally symmetric about every axis.

Now let's consider some Poisson Brackets involving angular momentum. For example, consider the PB's of x, y, and z with L_z:

$$\{x, L_z\} = \{x, (x\, p_y - y\, p_x)\}$$
$$\{y, L_z\} = \{y, (x\, p_y - y\, p_x)\} \qquad (19)$$
$$\{z, L_z\} = \{z, (x\, p_y - y\, p_x)\}.$$

You can work out these PB's using the definition Eq. (1), or you can use the axioms.

Exercise 3: Using both the definition of PB's and the axioms, work out the PB's in Equations (19). *Hint: In each expression, look for things in the parentheses that have nonzero Poisson Brackets with the coordinate x, y, or z. For example, in the first PB, x has a nonzero PB with p_x.*

Here are the results:

$$\{x, L_z\} = -y$$
$$\{y, L_z\} = x$$
$$\{z, L_z\} = 0.$$

If we compare this with Equations (18) we see a very interesting pattern. By taking the PB's of the coordinates with L_z we reproduce (apart from the ϵ) the expressions for the infinitesimal rotation about the z axis. In other words,

$$\{x, L_z\} \sim \delta x$$
$$\{y, L_z\} \sim \delta y$$
$$\{z, L_z\} \sim \delta z.$$

where $\sim$ means "apart from the factor ϵ."

The fact that taking a PB with a conserved quantity gives the transformation behavior of the coordinates under a symmetry—the symmetry related to the conservation law—is not an accident. It is very general and gives us another way to think about the relationship between symmetry and conservation. Before we pursue this relationship further, let's explore other PB's involving angular momentum. First of all, it is easy to generalize to other components of L. Again, you can do it by cycling $x \to y$, $y \to z$, $z \to x$. You'll get six more equations, and you might wonder whether there is a nice way to summarize them. In fact there is.

Mathematical Interlude—The Levi-Civita Symbol

A good notation can be worth a lot of symbols, especially if it appears over and over. An example is the Kronecker delta symbol δ_{ij}. In this section I will give you another one, the *Levi-Civita symbol*, which is also called the ϵ symbol ϵ_{ijk}. As in the Kronecker case, the indices i, j, k represent the three directions of space, either x, y, z or 1, 2, 3. The Kronecker symbol takes on two values: either 1 or 0, depending on whether $i = j$ or $i \neq j$. The ϵ symbol takes on one of three values: 0, 1, or –1. The rules for ϵ_{ijk} are a little more complicated than those for δ_{ij}.

First of all, $\epsilon_{ijk} = 0$ if any two indices are the same.—for example, ϵ_{111} and ϵ_{223} are both zero. The only time ϵ_{ijk} is not zero is when all three indices are different. There are six possibilities: ϵ_{123}, ϵ_{231}, ϵ_{312}, ϵ_{213}, ϵ_{132}, ϵ_{321}. The first three have value 1, and the second three have value -1.

What is the difference between the two cases? Here is one way to describe it: Arrange the three numbers 1, 2, 3 on a circle, like a clock with only three hours (see Figure 1).

Figure 1: A circular arrangement of the numbers 1, 2, and 3.

Start at any of the three numbers and go around clockwise. You get (123), (231), or (312), depending on where you start. If you do the same going counterclockwise, you get (132), (213), or (321). The rule for the Levi-Civita symbol is that $\epsilon_{ijk} = 1$, for the clockwise sequences, and $\epsilon_{ijk} = -1$ for the counterclockwise sequences.

Back to Angular Momentum

Now, with the aid of the ϵ symbol, we can write the PB's for all the coordinates and all the components of $\vec{L}$:

$$\{x_i, L_j\} = \sum_k \epsilon_{ijk} x_k. \tag{20}$$

For example, suppose that you want to know $\{y, L_x\}$. Identifying 1, 2, 3 with x, y, z and plugging these into Eq. (20) we get

$$\{x_2, L_1\} = \epsilon_{213} x_3.$$

Since 213 is a counterclockwise sequence, $\epsilon_{213} = -1$, so

$$\{x_2, L_1\} = -x_3.$$

Let's consider another set of PB's—namely, the PB's of p_i with the components of $\vec{L}$. They are easy to work out, and with the aid of the ϵ symbol, we get

$$\{p_i, L_j\} = \epsilon_{ijk}\, p_k.$$

For example,

$$\{p_x, L_z\} = -p_y.$$

The thing to notice is that the PB's of the p's and L's have exactly the same form as those of the x's and L's. That is interesting because the p's and x's transform exactly the same way under a rotation of coordinates. Just as $\delta x \sim - y$ for a rotation about z, the variation of p_x is proportional to $-p_y$.

The meaning of this is quite deep. It says that to compute the change in any quantity when the coordinates are rotated, we compute the Poisson bracket of the quantity with the angular momentum. For a rotation about the ith axis,

$$\delta F = \{F, L_i\}. \tag{21}$$

The angular momentum is the *generator* of rotations.

We will come back to this theme, and to the intimate relationship connecting symmetry transformations, Poisson Brackets, and conserved quantities, but first I want to explain how PB's can be useful in formulating and solving problems.

Rotors and Precession

One thing we haven't done yet is to compute the PB's between different components of the angular momentum. The PB of anything with itself is always zero, but the PB of one component of $\vec{L}$ with another is not zero. Consider

$$\{L_x, L_y\} = \{(y\, p_z - z\, p_y), (z\, p_x - x\, p_z)\}.$$

Either by using the definition of PB's or by using the axioms, we will get

$$\{L_x, L_y\} = L_z.$$

Try it.

The general relation can be read off by cycling through x, y, z. Here it is using the Levi-Civita symbol:

$$\{L_i, L_j\} = \sum_k \epsilon_{ijk} L_k. \tag{22}$$

That's very pretty, but what can we do with it? To illustrate the power of relations such as Eq. (22), let's consider a small, rapidly spinning ball in outer space. Call it a *rotor*. At any instant there is an axis of rotation, and the angular momentum is along that axis. If the rotor is isolated from all influences, then its angular momentum will be conserved, and the axis of rotation will not change.

Now suppose the rotor has some electric charge. Because the rotor is rapidly spinning, it behaves like an electromagnet with its north and south poles along the rotation axis. The strength of the dipole is proportional to the rate of rotation—or, better yet—to the angular momentum. This won't make any difference unless we put the whole thing in a magnetic field $\vec{B}$. In that case, there will be some energy associated with any misalignment between $\vec{L}$ and $\vec{B}$ (see Figure 2).

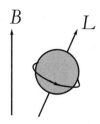

Figure 2: A rotor aligned at an angle to
a magnetic field.

That energy is proportional to the cosine of the angle
between the two vectors and to the product of their magnitudes.
In other words, the alignment energy is proportional to the dot
product

$$H \sim \vec{B} \cdot \vec{L}. \tag{23}$$

I've used the notation H for energy because later we will identify
it with the Hamiltonian of the system.

Let's take the magnetic field to be along the z axis so that
H is proportional to the z component of $\vec{L}$. Lumping the
magnetic field, the electric charge, the radius of the sphere, and
all the other unspecified constants into a single constant ω, the
energy of alignment takes the form

$$H = \omega L_z. \tag{24}$$

Let's pause for some perspective on what we are doing and
where we are going. It's obvious that without the magnetic field,
the system is rotationally symmetric in the sense that the energy
does not change if you rotate the axis of the rotor. But with the
magnetic field, there is something to rotate relative to. Therefore,
the rotational symmetry is ruined. Eq.s (23) and (24) represent

the rotational asymmetry. But what is the effect? The answer is obvious: The angular momentum is no longer conserved—no symmetry, no conservation. That means the direction of the spin will change with time, but exactly how?

One can try to guess the answer. The rotor is a magnet—like a compass needle—and intuition suggests that the angular momentum will swing toward the direction of $\vec{B}$, like a pendulum. That's wrong if the spin is very rapid. What does happen is that the angular momentum precesses, exactly like a gyroscope, around the magnetic field. (A gyroscope would precess about the gravitational field.) To see that, let's use the Poisson Bracket formulation of mechanics to work out the equations of motion for the vector $\vec{L}$.

First, recall that the time derivative of any quantity is the PB of that quantity with the Hamiltonian. Applying this rule to the components of $\vec{L}$ gives

$$\dot{L}_z = \{L_z, H\}$$
$$\dot{L}_x = \{L_x, H\}$$
$$\dot{L}_y = \{L_y, H\}.$$

or, using Eq. (24)

$$\dot{L}_z = \omega \{L_z, L_z\}$$
$$\dot{L}_x = \omega \{L_x, L_z\}$$
$$\dot{L}_y = \omega \{L_y, L_z\}.$$

Now we can see the point. Even if we know nothing about the material that the rotor is made of, where the charge resides, or how many particles are involved, we can solve the problem: We know the PB's between all components of $\vec{L}$. First we take the equation for L_z. Since it involves the PB of L_z with itself,

$$\dot{L}_z = 0.$$

The z component of $\vec{L}$ does not change. That immediately precludes the idea that $\vec{L}$ swings like a pendulum about $\vec{B}$.

Next we use Eq. (22) to work out $\dot{L}_x$ and $\dot{L}_y$:

$$\dot{L}_x = -\omega L_y$$

$$\dot{L}_y = \omega L_x.$$

This is exactly the equation of a vector in the x, y plane rotating uniformly about the origin with angular frequency ω. In other words, $\vec{L}$ precesses about the magnetic field. The magic of Poisson Brackets allows us to solve the problem knowing very little other than that the Hamiltonian is proportional to $\vec{B} \cdot \vec{L}$.

Symmetry and Conservation

Let's go back to Eq. (21), the meaning of which is that the variation of any quantity, under the action of a rotation, is proportional to the PB of that quantity with L_i. Moreover, L_i happens to be the quantity that is conserved by virtue of invariance with respect to rotation. That's an interesting connection, and one wonders how general it is. Let me give a couple of other examples of the same thing. Consider a particle on a line. If there is translation invariance, then the momentum p is conserved. Now take the PB of any function of x with p:

$$\{F(x), p\} = \frac{d\,F}{d\,x}.$$

What is the change in $F(x)$ under an infinitesimal translation by distance ϵ? The answer is

$$\delta F = \epsilon \frac{d\,F}{d\,x},$$

or

$$\delta F = \epsilon \{F(x), p\}.$$

Here's another example: If a system has time-translation invariance, then the Hamiltonian is conserved. What is the small change in a quantity under a time translation? You guessed it— the time derivative of the PB of the quantity with H.

Let's see if we can generalize the connection. Let $G(q, p)$ be any function of the coordinates and momenta of a system. I use the letter G because I am going to call it a *generator*. What it generates is small displacements of the phase space points. By definition, we will shift every point in phase space by the amount $\delta\,q_i, \delta\,p_i$, where

$$\delta\,q_i = \{q_i, G\}$$
$$\delta\,p_i = \{p_i, G\}. \tag{25}$$

Equations (25) generate an infinitesimal transformation of phase space. The transformation generated by G may or may not be a symmetry of the system. What exactly does it mean to say that it is a symmetry? It means that no matter where you start, the transformation does not change the energy. In other words, if $\delta H = 0$ under the transformation generated by G, then the transformation is a symmetry. We can therefore write that the condition for a symmetry is

$$\{H, G\} = 0. \tag{26}$$

But Eq. (26) can be read another way. Since interchanging the order of the two functions in a PB changes only the sign, Eq. (26) may be expressed as

$$\{G, H\} = 0. \tag{27}$$

which is exactly the condition that G is conserved. One can say it this way: The same Poisson Bracket that tells us how H changes under the transformation generated by G also tells us how G changes with time.

Lecture 11: Electric and Magnetic Forces

He kept a magnet in his coat pocket. How it attracted nails and other bits of metal was an endless source of fascination, and the way it spun the needle of his compass, round and round the world. What magic was inside that horseshoe-shaped bit of iron? Whatever it was, Lenny never tired of playing with his favorite toy.

What he didn't know was that the whole Earth is a magnet. Or that the earth-magnet was a providential force, that protected him from deadly solar radiation, bending the paths of charged particles into safe orbits. For the moment such things were beyond Lenny's imagination.

"Tell me about magnets, George."

Vector Fields

A field is nothing but a function of space and time that usually represents some physical quantity that can vary from point to point and from time to time. Two examples taken from meteorology are the local temperature and the air pressure. Since the temperature can vary, it makes sense to think of it as a function of space and time, $T(x, y, z, t)$, or, more simply, $T(x, t)$. The temperature and air pressure are obviously not vector fields. They have no sense of direction, nor do they have components in different directions. Asking for the y component of temperature is nonsense. A field that consists of only one number at each point of space is called a *scalar field*. The temperature field is a scalar.

There are, however, vector fields such as the local wind velocity. It has a magnitude, a direction, and components. We can write it as $\vec{v}(x, t)$, or we can write its components $v_i(x, t)$. Other examples of vector fields are the electric and magnetic fields created by electric charges and currents.

Because such fields vary in space, we can construct new fields by differentiating the original fields. For example, the three partial derivatives of temperature, $\frac{\partial T}{\partial x}, \frac{\partial T}{\partial y}, \frac{\partial T}{\partial z}$, can be considered to be the components of a vector field called the temperature *gradient*. If the temperature increases from north to south then the gradient points toward the south. Let's spend a little time going over the tricks used to create new fields from old ones by differentiating.

Mathematical Interlude: Del

Let's invent a fake vector called $\vec{\nabla}$. The verbal name of ∇ is "del," standing, I suppose, for delta, although an honest delta is written as Δ. The components of $\vec{\nabla}$ are not numbers. They are derivative symbols:

$$\nabla_x \equiv \frac{\partial}{\partial x}$$

$$\nabla_y \equiv \frac{\partial}{\partial y} \tag{1}$$

$$\nabla_z \equiv \frac{\partial}{\partial z}$$

At first sight Equations (1) look like nonsense. The components of vectors are numbers, not derivative symbols. And anyway the derivative symbols don't make sense—derivatives of what? The point is that ∇ never stands alone. Just like the derivative symbol $\dfrac{d}{dx}$, it must act on something—it must have a function of some sort to differentiate. For example, ∇ can act on a scalar such as the temperature. The components of $\nabla\, T$ are

$$\nabla_x T \equiv \frac{\partial T}{\partial x}$$

$$\nabla_y T \equiv \frac{\partial T}{\partial y}$$

$$\nabla_z T \equiv \frac{\partial T}{\partial z}.$$

and they indeed form the components of a genuine vector field— the gradient of the temperature. In a similar way, we can form the gradient of any scalar field.

Next, let's define the *divergence* of a vector field. The divergence is defined in analogy with the dot product of two vectors $\vec{V} \cdot \vec{A} = V_x A_x + V_y A_y + V_z A_z$, which, by the way, is a scalar. The divergence of a vector is also a scalar. Let the vector field be $\vec{A}(x)$. The divergence of $\vec{A}$ is the dot product of $\vec{\nabla}$ and $\vec{A}$—in other words, $\vec{\nabla} \cdot \vec{A}$. The meaning of this symbol is easy to guess by analogy with the usual dot product:

$$\vec{\nabla} \cdot \vec{A} = \frac{\partial A_x}{\partial x} + \frac{\partial A_y}{\partial y} + \frac{\partial A_z}{\partial z}. \tag{2}$$

Then consider the cross product of two vectors $\vec{V}$ and $\vec{A}$ which gives another vector. The components of the cross product are

$$\left(\vec{V} \times \vec{A}\right)_x = V_y A_z - V_z A_y$$

$$\left(\vec{V} \times \vec{A}\right)_y = V_z A_x - V_x A_z$$

$$\left(\vec{V} \times \vec{A}\right)_z = V_x A_y - V_y A_z.$$

Here is another way to write them using the Levi-Civita symbol:

$$\left(\vec{V} \times \vec{A}\right)_i = \sum_k \sum_j \epsilon_{ijk} V_j A_k. \tag{3}$$

Exercise 1: Confirm Eq. (3). Also prove that

$$V_i A_j - V_j A_i = \sum_k \epsilon_{ijk} \left(\vec{V} \times \vec{A}\right)_i.$$

Now let's substitute the fake vector $\vec{\nabla}$ for $\vec{V}$ in Eq. (3):

$$\left(\vec{\nabla} \times \vec{A}\right)_i = \sum_k \sum_j \epsilon_{ijk} \frac{\partial A_k}{\partial x_j}.$$

More explicitly

$$\left(\vec{\nabla} \times \vec{A}\right)_x = \frac{\partial A_z}{\partial y} - \frac{\partial A_y}{\partial z}$$

$$\left(\vec{\nabla} \times \vec{A}\right)_y = \frac{\partial A_x}{\partial z} - \frac{\partial A_z}{\partial x}$$

$$\left(\vec{\nabla} \times \vec{A}\right)_z = \frac{\partial A_y}{\partial x} - \frac{\partial A_x}{\partial y}.$$

What we have done is to start with a vector field $\vec{A}(x)$ and generate another vector field $\vec{\nabla} \times \vec{A}$ by differentiating A in a particular way. The new vector field $\vec{\nabla} \times \vec{A}$ is called the *curl* of $\vec{A}$.

Here is a theorem that takes a few seconds to prove. For any starting field $\vec{A}(x)$, the curl of A has no divergence,

$$\vec{\nabla} \cdot \left[\vec{\nabla} \times \vec{A}\right] = 0.$$

The theorem actually has a stronger form that is harder to prove. A field has zero divergence if and only if it is the curl of another field.

Here is another theorem that is easy to prove. Let a vector field be defined by the gradient of a scalar field:

$$\vec{E}(x) = \vec{\nabla} V(x)$$

where V is the scalar. Then it follows that the curl of $\vec{E}$ is zero:

$$\vec{\nabla} \times \left[\vec{\nabla} V(x)\right] = 0. \tag{4}$$

Exercise 2: Prove Eq. (4).

Magnetic Fields

Magnetic fields (called $\vec{B}(x)$) are vector fields, but not just any vector field can represent a magnetic field. All magnetic fields have one characteristic feature: Their divergence is zero. Thus it follows that any magnetic field can be expressed as a curl of some auxiliary field:

$$\vec{B} = \vec{\nabla} \times \vec{A} \qquad (5)$$

where $\vec{A}$ is called the *vector potential*. In component form,

$$B_x = \frac{\partial A_z}{\partial y} - \frac{\partial A_y}{\partial z}$$

$$B_y = \frac{\partial A_x}{\partial z} - \frac{\partial A_z}{\partial x} \qquad (6)$$

$$B_z = \frac{\partial A_y}{\partial x} - \frac{\partial A_x}{\partial y}.$$

The vector potential is a peculiar field. In a sense it does not have the same reality as magnetic or electric fields. It's only definition is that its curl is the magnetic field. A magnetic or electric field is something that you can detect locally. In other words, if you want to know whether there is an electric/magnetic field in a small region of space, you can do an experiment in that same region to find out. The experiment usually consists of seeing whether there are any forces exerted on charged particles in that region. But vector potentials cannot be detected locally. First of all, they are not uniquely defined by the magnetic field they are representing. Suppose $\vec{B}$ is given by a vector potential, as in Eq. (5). We can always add a gradient to $\vec{A}$ to define a new vector potential without changing $\vec{B}$. The reason is that the curl of a gradient is always zero. Thus if two vector potentials are related by

$$\vec{A'} = \vec{A} + \vec{\nabla} s$$

for some scalar s, then they produce identical magnetic fields and cannot be distinguished by any experiment.

This is not the first time we have seen an ambiguity having to do with one thing being defined by a derivative of another. Remember that the force on a system is minus the gradient of the potential energy:

$$\vec{F}(x) = -\vec{\nabla} U(x).$$

The potential energy is not unique: You can always add a constant without changing the force. This means that you can never directly measure the potential, but only its derivative. The situation is similar with the vector potential; indeed, that's why it is called a potential.

Let's consider an example of a magnetic field and its associated vector potential. The simplest case is a uniform magnetic field pointing, say, along the z axis:

$$B_x = 0$$
$$B_y = 0 \tag{7}$$
$$B_z = b,$$

where b is a number representing the strength of the field. Now define a vector potential by

$$A_x = 0$$
$$A_y = b x \tag{8}$$
$$A_z = 0.$$

When the curl of $\vec{A}$ is computed, there is only one term, namely

$\dfrac{\partial A_y}{\partial x} = b$. Thus the only component of the magnetic field is the z component, and it has value b.

Now, there is something funny about Equations (8). The uniform magnetic field seems to be completely symmetric with respect to rotations in the x, y plane. But the vector potential has only a y component. However, we could have used a different vector potential $\vec{A'}$—one with only an x component—to generate the very same magnetic field:

$$A'_x = -b\,y$$
$$A'_y = 0 \qquad\qquad (9)$$
$$A'_z = 0.$$

Exercise 3: Show that the vector potentials in Equations (8) and Equations (9) both give the same uniform magnetic field. This means that the two differ by a gradient. Find the scalar whose gradient, when added to Equations (8), gives Equations (9).

The operation of changing from one vector potential to another to describe the same magnetic field has a name. It is called a *gauge transformation*. Why "gauge"? It's a historical glitch. At one time it was wrongly thought to reflect ambiguities in gauging lengths at different locations.

If the vector potential is ambiguous but the magnetic field quite definite, why bother with the vector potential at all? The answer is that without it, we could not express the principle of stationary action, or the Lagrangian, Hamiltonian, and Poisson formulations of mechanics for particles in magnetic fields. It's a weird situation: The physical facts are *gauge invariant*, but the formalism requires us to choose a gauge (a particular choice of vector potential).

The Force on a Charged Particle

Electrically charged particles are influenced by electric and magnetic fields $\vec{E}$ and $\vec{B}$. The force due to the electric field is simple and of the form that we studied in earlier chapters; specifically, it is the gradient of a potential energy. In terms of the electric field,

$$\vec{F} = e\vec{E},$$

where e is the charge of the particle. It is a rule of electromagnetic theory that a static (not time-dependent) electric field has no curl so it must be a gradient. The usual notation is

$$\vec{E} = -\vec{\nabla} V,$$

so we can write the force as

$$\vec{F} = -e\vec{\nabla} V.$$

The potential energy is $e\vec{\nabla} V$, and everything is completely conventional.

Magnetic forces on charged particles are different and a little more complicated. They depend not only on the position of the particle through the value of the magnetic field, but also on the velocity of the particle. They are referred to as *velocity-dependent forces*. The magnetic force on a charged particle was first written down by the great Dutch physicist H. A. Lorentz and is called the

Lorentz force. It involves the velocity vector of the particle and the speed of light c:

$$\vec{F} = \frac{e}{c}\,\vec{v} \times \vec{B}. \tag{10}$$

Notice that the Lorentz force is perpendicular to both the velocity and the magnetic field. Combining Eq. (10) with Newton's $\vec{F} = m\,\vec{a}$, we find that the equations of motion for a particle in a magnetic field are

$$m\,\vec{a} = \frac{e}{c}\,\vec{v} \times \vec{B}. \tag{11}$$

The Lorentz force is not the first velocity-dependent force we have encountered. Recall that in a rotating frame, there are two so-called fictitious forces: The centrifugal force and (more to the point) the Coriolis force. The Coriolis force is given by

$$\vec{F} = 2\,m\,\vec{v} \times \vec{\omega}, \tag{12}$$

where $\vec{\omega}$ is the vector representing the angular velocity of the rotating frame. The Coriolis and Lorentz forces are very similar, with the magnetic field and the angular velocity playing the same role. Of course, not all magnetic fields are uniform, so the magnetic situation can be far more complex than the Coriolis case.

The Lagrangian

All of this raises the question of how to express magnetic forces in the action, or Lagrangian, form of mechanics. One source of confusion is that the symbol for action and the symbol for vector potential are both A. In what follows, we will use A for the action, and $\vec{A}$, or A_i, as the vector potential. Let's ignore or set equal to zero the electric field and concentrate on the magnetic, or Lorentz force. Begin with the action for a free particle with no forces:

$$A = \int_{t_0}^{t_1} L\left(x, \dot{x}\right) d t$$

with

$$L = \frac{m}{2}\left(\dot{x}_i\right)^2.$$

Here i refers to the direction of space, and the summation sign for summing over x, y, z has been left implicit. Get used to it.

What can we add to the action or to the Lagrangian to give rise to a Lorentz force? The answer is not obvious. However, we know that whatever the additional ingredient is, it should be proportional to the electric charge, and it should also involve the magnetic field in some form.

You can experiment around with it and get frustrated. There is nothing you can do directly involving $\vec{B}$ that will give the Lorentz force. The key is the vector potential. The simplest thing we can do with the vector potential is to dot it into the velocity vector. Remember that the Lagrangian involves only the positions and the velocities. You might also try dot products of the position vector with $\vec{A}$, but that doesn't work very well. So let's try adding to the Lagrangian the term

$$\frac{e}{c}\,\vec{v}\cdot\vec{A}(x) = \frac{e}{c}\sum_i\left[\dot{x}_i\,A_i(x)\right]. \tag{13}$$

The reason for including the speed of light is that it occurs together with the charge in Lorentz force. Thus we try out the action:

$$A = \int_{t_0}^{t_1}\sum_i\left[\frac{m}{2}\left(\dot{x}_i\right)^2 + \frac{e}{c}\,\dot{x}_i\cdot A_i(x)\right]d\,t. \tag{14}$$

Now you might object that the equation of motion is not supposed to involve the vector potential, but only the magnetic field. We know that the vector potential is not unique, so won't we get another answer if we make a gauge transformation $\vec{A}' = \vec{A} + \vec{\nabla}\,s$? Let's see what happens to the action if we do so.

The important part of the action is the term arising from Eq. (13):

$$A_L = \frac{e}{c}\int_{t_0}^{t_1}\sum_i\left[\dot{x}_i\,A_i(x)\right]d\,t.$$

or, more explicitly,

$$A_L = \frac{e}{c}\int_{t_0}^{t_1}\sum_i\left[A_i(x)\,\frac{d\,x_i}{d\,t}\right]d\,t.$$

In this equation, A_L is the part of the action that we are adding to try to account for the Lorentz force—hence the subscript L. Suppose we change $\vec{A}$ by adding $\vec{\nabla}\,s$. At first sight, it would seem to change A_L by adding the term

$$\frac{e}{c} \int_{t_0}^{t_1} \sum_i \left(\frac{\partial s}{\partial x_i} \frac{d\,x_i}{d\,t} \right) d\,t.$$

If you look at this carefully, you will see that it all boils down to a simple expression. The $d\,t$'s in the numerator and denominator cancel:

$$\frac{e}{c} \sum_i \left(\int_{t_0}^{t_1} \frac{\partial s}{\partial x_i} d\,x_i \right).$$

And then the whole thing is just the difference between the value of s at the beginning and its value at the end of the trajectory. In other words, the gauge transformation added a term $s_1 - s_0$ to the action, where s_0 and s_1 are the values of s at the initial and final positions of the trajectory, respectively. In other words, the change in the action due to the gauge transformation is

$$s_1 - s_0. \tag{15}$$

Does such a change make any difference to the equations of motion? Let's recall exactly what the action principle actually says. Given any two points in space and time, x_0, t_0 and x_1, t_1, there are many trajectories that connect them, but only one is the true trajectory taken by a particle. The true trajectory is the one that minimizes, or makes stationary, the action. So what we do is explore all trajectories that connect the points until we find the stationary-action solution. From that principle we derived the Euler-Lagrange equations of motion.

As we see in Eq. (15), a gauge transformation changes the action, but only if we vary the endpoints. If the endpoints are kept fixed, the change in the action has no effect. The stationary point has to do only with changing the trajectory without moving the endpoints. Though the action changes, the equations of motion do not, and neither do the solutions. We say that the equations of motion and their solutions are *gauge invariant*.

One more bit of jargon: Since there are many possible choices of vector potentials that describe the same physical situation, a specific choice is simply called a *gauge*. For example, Equations (8) and Equations (9) are two different gauges describing the same uniform magnetic field. The physical principle that the result of any experiment should not depend on the gauge choice is called *gauge invariance*.

Equations of Motion

Let's return to the action, Eq. (14). And let's be very explicit about the Lagrangian:

$$L = \frac{m}{2}\left(\dot{x}^2 + \dot{y}^2 + \dot{z}^2\right) + \frac{e}{c}\left(\dot{x}\,A_x + \dot{y}\,A_y + \dot{z}\,A_z\right). \qquad (16)$$

Starting with x, the Lagrange equation of motion is

$$\dot{p}_x = \frac{\partial L}{\partial x}. \qquad (17)$$

First the canonical momenta: You might think that the momenta are just the usual mass times velocity, but that's not right. The correct definition is that the momenta are the derivatives of the Lagrangian with respect to the components of velocity. This does give $p = mv$ with the usual particle Lagrangians, but not with a magnetic field. From Eq. (16) we get

$$p_x = m\dot{x} + \frac{e}{c}\,A_x. \qquad (18)$$

This may worry you. It indicates that the canonical momentum is not gauge invariant. This is true, but we are not finished yet. We have two more things to do. We must compute the time derivative of p_x and also compute the right-hand side of Eq. (17). Maybe, if we are lucky, all the gauge-dependent stuff will cancel.

The left-hand side of Eq. (17) is

$$\dot{p}_x = m\, a_x + \frac{e}{c}\frac{d}{dt}A_x$$

$$= m\, a_x + \frac{e}{c}\left(\frac{\partial A_x}{\partial x}\dot{x} + \frac{\partial A_x}{\partial y}\dot{y} + \frac{\partial A_x}{\partial z}\dot{z}\right),$$

where a_x is the x component of acceleration.

The right-hand side side of Eq. (17) is:

$$\frac{\partial L}{\partial x} = \frac{e}{c}\left(\frac{\partial A_x}{\partial x}\dot{x} + \frac{\partial A_y}{\partial x}\dot{y} + \frac{\partial A_z}{\partial x}\dot{z}\right).$$

Now let's combine the left and right sides:

$$m\, a_x = \frac{e}{c}\left(\frac{\partial A_y}{\partial x} - \frac{\partial A_x}{\partial y}\right)\dot{y} + \frac{e}{c}\left(\frac{\partial A_z}{\partial x} - \frac{\partial A_x}{\partial z}\right)\dot{z}. \qquad (19)$$

Equation (19) looks complicated, but note that the combinations of derivatives

$$\frac{\partial A_y}{\partial x} - \frac{\partial A_x}{\partial y}$$

and

$$\frac{\partial A_z}{\partial x} - \frac{\partial A_x}{\partial z}$$

are things we saw in Equations (7)—namely, the z and y

components of the magnetic field. We can rewrite Eq. (19) in a much simpler form:

$$m\,a_x = \frac{e}{c}\left(B_z\,\dot{y} - B_y\,\dot{z}\right).\tag{20}$$

Take a careful look at Eq. (20). You should be impressed by a number of things. First of all, the equation is gauge invariant: On the right-hand side, the vector potential has completely disappeared in favor of the magnetic field. The left-hand side is the mass times the acceleration—that is the left-hand side of Newton's equation. In fact, Eq. (20) is nothing but the x component of the Newton-Lorentz equation of motion, Eq. (12).

One might wonder why we bothered introducing the vector potential at all. Why not just write the gauge-invariant Newton-Lorentz equation? The answer is that we can, but then we lose any possibility of formulating the equations as an action principle, or as Hamilton's equations of motion. That might not be such a tragedy for the classical theory, but it would be a disaster for quantum mechanics.

The Hamiltonian

Before discussing the Hamiltonian of a charged particle in a magnetic field, let's go back to the definition of the particle's momentum. You may still find it confusing. The reason is that there are two separate concepts: mechanical momentum and canonical momentum. Mechanical momentum is what you learn about in elementary mechanics (*Momentum equals mass times velocity*) and in advanced mechanics (*Canonical momentum equals derivative of the Lagrangian with respect to velocity*). In the simplest situations where the Lagrangian is just the difference of kinetic and potential energy, the two kinds of momentum are the same. That's because the only dependence on velocity is $\frac{1}{2} m v^2$.

But once the Lagrangian gets more complicated, the two kinds of momentum may not be the same. In Eq. (18) we see such an example. The canonical momentum is the mechanical momentum plus a term proportional to the vector potential. We can write it in vector notation:

$$\vec{p} = m \vec{v} + \frac{e}{c} \vec{A}(x). \tag{21}$$

The mechanical momentum is not only familiar; it is gauge invariant. It is directly observable, and in that sense it is "real." The canonical momentum is unfamiliar and less "real"; it changes when you make a gauge transformation. But whether or not it is real, it is necessary if you want to express the mechanics of charged particles in Lagrangian and Hamiltonian language.

To pass to the Hamiltonian, we recall the definition

$$H = \sum_i \left(p_i \, \dot{q}_i \right) - L,$$

which in this case is

$$H = \sum_i \left\{ p_i \dot{x}_i - \left[\frac{m}{2} \left(\dot{x}_i \right)^2 + \frac{e}{c} \dot{x}_i \cdot A_i(x) \right] \right\}. \tag{22}$$

Let's work it out. First we will need to get rid of the velocities; the Hamiltonian is always thought of as a function of coordinates and momenta. That's easy. We just solve Eq. (21) for velocity in terms of p:

$$\dot{x}_i = \frac{1}{m}\left[p_i - \frac{e}{c}A_i(x)\right].$$ (23)

Now wherever you see a velocity component in Eq. (22), substitute Eq. (23) and then do a little rearranging. Here is what you will get:

$$H = \sum_i \left\{\frac{1}{2m}\left[p_i - \frac{e}{c}A_i(x)\right]\left[p_i - \frac{e}{c}A_i(x)\right]\right\}.$$ (24)

Exercise 4: Using the Hamiltonian, Eq. (24), work out Hamilton's equations of motion and show that you just get back the Newton-Lorentz equation of motion.

If you carefully look at Eq. (24), you will see something a little surprising. The combination $\left[p_i - \frac{e}{c}A_i(x)\right]$ is the mechanical momentum $m\,v_i$. The Hamiltonian is nothing but

$$H = \frac{1}{2}m\,v^2.$$

In other words, its numerical value is the same as the naive kinetic energy. That proves (among other things) that the energy is gauge invariant. Since it conserved, the naive kinetic energy is also conserved, as long as the magnetic field does not change with time. But that does not mean the particle motion does not sense the magnetic field. If you want to use the Hamiltonian to find the motion, you must express it in terms of the canonical momentum, not the velocity, and then use Hamilton's equations. Alternatively, you can work with velocities and use the Lagrangian form of the equations, but in that case the Lagrangian is not the naive kinetic energy. In either case, if you work it all out, you will discover that the charged particle experiences a gauge-invariant Lorentz magnetic force.

Motion in a Uniform Magnetic Field

Motion in a uniform magnetic field is easy enough to solve, and it illustrates a lot of the principles we have been discussing. Let's take the field to lie in the z direction and to have magnitude b. This is the situation described in Equations (6, 7, 8). The choice between the vector potentials in Equations (7, 8) is an example of the ambiguity associated with gauge transformations. Let's first choose Equations (7) and write the Hamiltonian, Eq. (24), using $\left(A_x = 0, A_y = b\,x, A_z = 0\right)$. We get

$$H = \frac{1}{2\,m}\left[(p_x)^2 + (p_z)^2 + \left(p_y - \frac{e}{c}\,b\,x\right)^2\right].$$

As always, the first thing to do is to look for conservation laws. We already know one: energy conservation. As we've seen, the energy is the old-fashioned kinetic energy $\frac{1}{2}\,m\,v^2$. It follows that the magnitude of the velocity is constant.

Next, notice that the only coordinate that appears in H is x. This means that when we work out Hamilton's equations, we will find that p_x is not conserved but that both p_z and p_y are conserved. Let's see what the implications are. First the z component. Since $A_z = 0$, $p_z = m\,v_z$, and the conservation of p_z tells the familiar story that the z component of velocity is constant.

Next look at the conservation of p_y. This time p_y is not equal to $m\,v_y$ but, rather, to $m\,v_y + \frac{e}{c}\,b\,x$. The conservation of p_y

then tells us that

$$m\,a_y + \frac{e}{c}\,b\,v_x = 0,$$

or

$$a_y = -\frac{e\,b}{m\,c}\,v_x. \tag{25}$$

Notice that the conservation of p_y does *not* imply that the y component of velocity is conserved.

What about p_x? It does not seem to be conserved since H has explicit dependence on x. We could use Hamilton's equations to determine the x component of acceleration, but I'm going to do it another way. Instead of using Equations (8) I'm going to change the gauge in midstream and use Equations (7). Remember that the physical phenomena should not change. The new Hamiltonian that goes with Equations (7) is,

$$H = \frac{1}{2\,m}\left[\left(p_x + \frac{e}{c}\,b\,y\right)^2 + \left(p_y\right)^2 + \left(p_z\right)^2\right].$$

Now the Hamiltonian does not depend on x, which implies that p_x is conserved. How can that be? We previously showed the x-component of momentum p_x is conserved when we used Equations (8). The answer is that when we make a gauge transformation the components of p change. In the two cases, p_x does not have the same meaning.

Let us see the implication of p_x conservation in the new gauge. Using Equations (7) we find that $p_x = m\,v_x - \frac{e}{c}\,b\,y$. Thus the conservation of p_x is expressed as

$$a_x = \frac{e\,b}{m\,c}\,v_y. \tag{26}$$

By now you may have already realized that Eq. (25) and Eq. (26) are familiar. They are the Newton-Lorentz equations for motion in a uniform magnetic field.

Exercise 5: Show that in the x, y plane, the solution to Eq. (25) and the solution to Eq. (26) are a circular orbit with the center of the orbit being anywhere on the plane. Find the radius of the orbit in terms of the velocity.

Gauge Invariance

The reason why I left magnetic forces for the last lecture is that I want you to remember the lessons when, in future study, we come to quantum mechanics and field theory. Gauge fields and gauge invariance are not minor artifacts of writing the Lorentz force in Lagrangian form. They are the central guiding principles that underlie everything, from quantum electrodynamics to general relativity and beyond. They play a leading role in condensed matter physics—for example, in explaining all sorts of laboratory phenomena such as superconductivity. I will close these lectures on classical mechanics by reviewing the meaning of the gauge idea, but its real importance will become clear only in later lectures.

The simplest meaning of a gauge field—the vector potential is the most elementary example—is that it is an auxiliary device that is introduced to make sure certain constraints are satisfied. In the case of a magnetic field, not any $\vec{B}(x)$ is allowed.

The constraint is that $\vec{B}(x)$ should have no divergence:

$$\vec{\nabla} \cdot \vec{B} = 0$$

To ensure that, we write the magnetic field as the curl of something—$\vec{A}(x)$—because curls automatically have no divergence. It's a trick to avoid having to worry explicitly about the fact that $\vec{B}(x)$ is constrained.

But we soon discover that we cannot get along without $\vec{A}(x)$. There is no way to derive Lorentz's force law from a Lagrangian without the vector potential. That is a pattern: To write the equations of modern physics in either Lagrangian or Hamiltonian form, auxiliary gauge fields have to be introduced.

But they are also nonintuitive and abstract. Despite their being indispensable, you can change them without changing the physics. Such changes are called *gauge transformations,* and the fact that physical phenomenal do not change is called *gauge invariance.* Gauge fields cannot be "real," because we can change them without disturbing the gauge invariant physics. On the other hand, we cannot express the laws of physics without them.

I am not about to give you a sudden insight that will resolve this tension. I will just say that's the way it is: The laws of physics involve gauge fields, but objective phenomena are gauge invariant.

Good Bye for Now

We are now finished with classical mechanics. If you have followed along, you know the Theoretical Minimum—all you need to know about classical mechanics to move on to the next thing. See you in Quantum Mechanics!

Appendix 1: Central Forces and Planetary Orbits

Lenny stooped and peered through the eyepiece of the telescope. It was the first time he had ever done that. He saw the rings of Saturn and whistled at their beauty. "George, have you seen the rings?"

George nodded and said, "Yup, I seen 'em."

Lenny looked up and pressed his friend. "Where do they come from?"

George said, "It's like the Earth goin' round the Sun."

Lenny nodded. "How does it go around?"

The Central Force of Gravity

A central force field is a force that points toward a center—in other words, toward a point of space (see Figure 1). In addition, for a force to be a central force, the magnitude of the force must be the same in every direction.

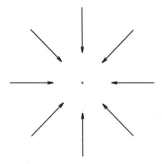

Figure 1: A central force.

Other than the obvious symmetry—rotational symmetry—there is nothing very special about central forces from a mathematical viewpoint. But their role in physics and in the history of physics is very special. The first problems solved by Newton—the problems of planetary orbits—were central force problems. The motion of an electron orbiting a hydrogen nucleus is a central force problem. Two atoms orbiting one another to form a simple molecule can be reduced to a central force problem in which the center is the center of mass. Since there was not enough time to cover this subject in the lectures, we'll add it here as a supplement.

Let's focus on the motion of the Earth as it orbits the much heavier Sun. According to Newton's laws, the force exerted by the Sun on the Earth is equal and opposite to that exerted by the Earth on the Sun. Moreover, the direction of those forces is along the line connecting the two bodies. Because the Sun is so much heavier than the Earth, the motion of the Sun is negligible, and it can be considered to be at a fixed location. We can choose our coordinates so that the Sun is at the origin, $x = y = z = 0$. The Earth, by contrast, moves in an orbit about the origin. Let's denote the location of the Earth by the vector $\vec{r}$ with components x, y, z. Since the Sun is located at the origin, the force on the Earth points toward the origin, as shown in Figure 1. Moreover, the magnitude of the force depends only on the distance r from the origin. A force with these properties— pointing toward the origin and depending only on the distance— is called a *central force*.

Let's rewrite the unit vector from Interlude 1:

$$\hat{r} = \frac{\vec{r}}{r}.$$

In equation form, the definition of a central force is

$$\vec{F} = f\left(\vec{r}\right)\hat{r},$$

where $f\left(\vec{r}\right)$ determines two things. First, the magnitude of $f\left(\vec{r}\right)$ is the magnitude of the force when the Earth is at distance r. Second, the sign of $f\left(\vec{r}\right)$ determines whether the force is toward or away from the Sun—in other words, whether the force is attractive or repulsive. In particular, if $f\left(\vec{r}\right)$ is positive the force is away from the Sun (repulsive), and if it is negative the force is toward the Sun (attractive).

The force between the Sun and the Earth is of course gravitational. According to Newton's law of gravitation, the gravitational force between two objects of mass m_1 and m_2 has the following properties.

N1: The force is attractive and proportional to the product of the objects' masses and a constant called G. Today we refer to G as Newton's constant. Its value is $G \approx 6.673 \text{ m}^3 \text{ kg}^{-1} \text{ s}^{-2}$.

N2: The force is inversely proportional the square of the distance between the masses.

To summarize, the force is attractive and has magnitude $\dfrac{G m_1 m_2}{r^2}$.

In other words, the function $f\left(\vec{r}\right)$ is given by

$$f\left(\vec{r}\right) = \frac{G m_1 m_2}{r^2},$$

and

$$\vec{F}_{\text{grav}} = -\frac{G m_1 m_2}{r^2}\, \hat{r}.$$

For the case of the Earth-Sun system let's denote the Sun's mass by M and the Earth's mass by m. The force on the Earth is

$$\vec{F}_{\text{grav}} = -\frac{G M m}{r^2}\, \hat{r}.$$

The equation of motion for the Earth's orbit is the usual $F = m\, a$, or, using the gravitational force,

$$m\, \frac{d^2 \vec{r}}{d t^2} = -\frac{G M m}{r^2}\, \hat{r}.$$

Notice an interesting fact: The mass of the Earth cancels from both sides of the equation, so the equation of motion does not depend on the mass of the Earth:

$$\frac{d^2 \vec{r}}{d t^2} = -\frac{G M}{r^2}\, \hat{r}. \tag{1}$$

An object of very different mass, such as a satellite, could orbit the Sun in the same orbit as the Earth. One caveat about this fact: It is true only if the Sun is so massive, compared with the Earth or satellite, that its motion can be ignored.

Gravitational Potential Energy

The gravitational force can be derived from a potential energy function. Recall that the force associated with a potential energy is minus the gradient of the potential:

$$F = -\nabla V.$$

It's not too hard to guess the form of V for the gravitational case. First of all, since the force is proportional to the constant $G M m$, one expects the potential energy also to have this factor.

Next, because the magnitude of the force only depends on the distance r, one may expect the potential energy $V(r)$ also to depend only on r. Finally, since we have to differentiate $V(r)$ to get the force, and since the force is proportional to $1 / r^2$, the potential energy must be proportional to $-1 / r$. Thus it is natural to try

$$V(r) = -\frac{G M m}{r}.$$

In fact, this is exactly right.

The Earth Moves in a Plane

Earlier, we mentioned that the central force problem has a symmetry. You probably recognize it as rotational symmetry about the origin. The implication of the symmetry, explained in Lecture 7, is the conservation of angular momentum. Suppose that at some instant the earth has location $\vec{r}$ and velocity $\vec{v}$. We can place these two vectors and the position of the Sun in a plane—the momentary plane of the Earth's orbit.

The angular momentum vector $\vec{L}$ is proportional to the cross product $\vec{r} \times \vec{v}$, so it is perpendicular to both $\vec{r}$ and $\vec{v}$ (see Figure 2). In other words, the angular momentum is perpendicular to the plane of the orbit. This is a powerful fact when combined with the conservation of angular momentum.

The conservation tells us that the vector $\vec{L}$ never changes. From that we conclude that the orbital plane never changes. To put it simply, the Earth's orbit and the Sun permanently lie in a fixed plane that does not vary. Knowing this, we may rotate our coordinates so that the orbit is in the x, y plane. The entire problem is then two-dimensional, the third coordinate z playing no role.

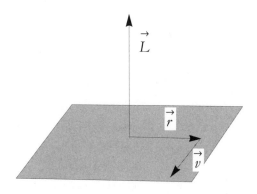

Figure 2: The relationship among the angular momentum $\vec{L}$, the position vector $\vec{r}$, and the velocity $\vec{v}$.

Polar Coordinates

We could work with the Cartesian coordinates x, y, but central force problems are much easier to solve in polar coordinates r, θ:

$$r = \sqrt{x^2 + y^2}$$

$$\cos \theta = \frac{x}{r}$$

In polar coordinates the kinetic energy of the earth is simple enough:

$$T = \frac{m}{2}\left(\dot{r}^2 + r^2\dot{\theta}^2\right). \tag{2}$$

The potential energy is even simpler—it does not involve θ at all:

$$V(r) = -\frac{GMm}{r}. \tag{3}$$

Equations of Motion

As is usually the case, the easiest route to the equations of motion is through the Lagrangian method. Recall that the Lagrangian is the difference of the kinetic and potential energies, $L = T - V$. Using Eq. (2) and Eq. (3), the Lagrangian in polar coordinates is

$$L = \frac{m}{2}\left(\dot{r}^2 + r^2\dot{\theta}^2\right) + \frac{GMm}{r}. \tag{4}$$

The equations of motion,

$$\frac{d}{dt}\frac{\partial L}{\partial \dot{r}} = \frac{\partial L}{\partial r}$$

$$\frac{d}{dt}\frac{\partial L}{\partial \dot{\theta}} = \frac{\partial L}{\partial \theta},$$

take the explicit form

$$\ddot{r} = r\dot{\theta} - \frac{GM}{r} \tag{5}$$

and

$$\frac{d}{dt}\left(m\,r^2\,\dot{\theta}\right) = 0. \qquad (6)$$

This last equation has the form of a conservation law. Not surprisingly, it is conservation of angular momentum. (To be precise, it is the conservation of the z component of angular momentum.) It is traditional to denote the angular momentum by the symbol L, but we are using that as the Lagrangian, so we will use p_θ instead. If we know p_θ at any particular instant, then we know it for all time. We may write

$$m\,r^2\,\dot{\theta} = p_\theta \qquad (7)$$

and just treat p_θ as a known constant.

This enables us to express the angular velocity in terms of the distance of the Earth from the Sun. We just solve the equation for $\dot{\theta}$:

$$\dot{\theta} = \frac{p_\theta}{m\,r^2}. \qquad (8)$$

We will come back to this relation between angular velocity and radial distance, but first let's return to the equation for r, namely

$$m\,\ddot{r} = m\,r\,\dot{\theta}^2 - \frac{G\,M\,m}{r^2}. \qquad (9)$$

In Eq. (9) the angular velocity appears, but we can use Eq. (8) to replace it:

$$m\,\ddot{r} = \frac{p_\theta^2}{m\,r^3} - \frac{G\,M\,m}{r^2}. \qquad (10)$$

The equation for r has an interesting interpretation. It looks like the equation for a single coordinate r under the influence of a combined "effective" force:

$$F_{\text{effective}} = \frac{p_\theta^2}{m\,r^3} - \frac{GMm}{r^2}. \tag{11}$$

The term $-\frac{GMm}{r^2}$ is just the gravitational force, but at first sight the second term may be a surprise. It is, in fact, nothing but the fictitious centrifugal force experienced by any particle that has an angular motion about the origin.

It's useful to pretend that Eq. (11) really does describe a particle moving under a total force that includes both the real gravitational force and the centrifugal force. Of course, for each value of the angular momentum, we must readjust p_θ, but since p_θ is conserved, we may regard it as a fixed number.

Given the effective force, one can also construct an effective potential energy function that includes the effect of gravity and the effect of centrifugal force:

$$V_{\text{effective}} = \frac{p_\theta^2}{2\,m\,r^2} - \frac{GMm}{r}. \tag{12}$$

You can easily check that

$$F_{\text{effective}} = -\frac{d\,V_{\text{effective}}}{d\,r}.$$

For all practical purposes, we can pretend the r motion is just that of a particle whose kinetic energy has the usual form, $\frac{m\,\dot{r}^2}{2}$, whose potential energy is $V_{\text{effective}}$, and whose Lagrangian is

$$L_{\text{effective}} = \frac{m \, \dot{r}^2}{2} - \frac{p_\theta^2}{2 \, m \, r^2} + \frac{G M m}{r}. \tag{13}$$

Effective Potential Energy Diagrams

In getting a feel for a problem, it is often a good idea to make a graph of the potential energy. For example, the equilibrium points (where the system may be at rest) can be identified as the stationary points (minima, maxima) of the potential. In understanding central force motion, we do exactly the same, except that we apply it to the effective potential. Let's first plot the two terms in $V_{\text{effective}}$ separately, as shown in Figure 3. Note that the two terms are of opposite sign; the centrifugal term is positive and the gravitational term negative. The reason is that the gravitational force is attractive, whereas the centrifugal force pushes the particle away from the origin.

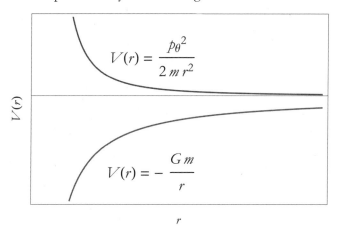

Figure 3: The potential energy diagram for the centrifugal and gravitational terms.

Near the origin the centrifugal term is the most

important, but at large values of r the gravitational term has the larger magnitude. When we combine them, we get a graph of $V_{\text{effective}}$ that looks like Figure 4.

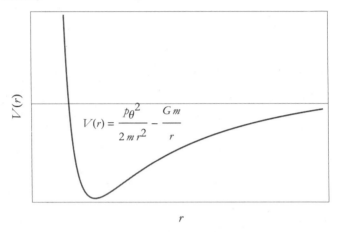

$$V(r) = \frac{p_\theta^{\,2}}{2\,m\,r^2} - \frac{G\,m}{r}$$

Figure 4: The potential energy diagram for the combination of centrifugal and gravitational terms.

Note that when the two terms are combined, the graph has a minimum. That may seem odd; we don't expect an equilibrium point where the Earth can stand still. But we have to remember that we are discussing only the behavior of r and ignoring the angular coordinate θ. The point is that for each angular momentum, there are orbits that maintain a constant radial distance while moving around the Sun. Such orbits are circular. On the graph of $V_{\text{effective}}$ a circular orbit is represented by a fictitious particle sitting at rest at the minimum.

Let's compute the value of r at the minimum. All we have to do is differentiate $V_{\text{effective}}$ and set the derivative equal to zero. It's an easy calculation that I will leave to you. The result is that the minimum occurs at

$$r = \frac{p_\theta^2}{GMm^2}.$$

(14)

Equation (14) yields the radius of the Earth's orbit (assuming it is circular, which is not quite right) given its angular momentum.

Kepler's Laws

Tycho Brahe was a sixteenth-century Danish astronomer before the age of telescopes. With the help of a long rod and some instruments to measure angles, he made the best tables and records of the motion of the Solar System before telescopes were invented. As a theoretician, he was somewhat confused. His legacy was his tables.

It was Tycho's assistant Johannes Kepler who put the tables to good use. Johannes Kepler took those records and fit the observed data to simple geometric and mathematical facts. He had no idea why the planets moved according to his laws—by modern standards, his theories of *why* were at best, odd—but he got the facts right.

Newton's great achievement—in a sense the start of modern physics—was to explain Kepler's laws of planetary motion through his own laws of motion, including the inverse-square law of gravity. Let's recall Kepler's three laws.

K1: The orbit of every planet is an ellipse with the Sun at one of the two foci.

K2: A line joining a planet and the Sun sweeps out equal areas during equal intervals of time.

K3: The square of the orbital period of a planet is directly proportional to the cube of the radius of its orbit.

Begin with K1, the ellipse law. Earlier we explained that circular orbits correspond to being in equilibrium at the minimum of the effective potential. But there are motions of the effective one-dimensional system in which it oscillates back and forth near, but not at, the minimum. A motion of this type would have the Earth periodically getting closer to and farther from the Sun. Meanwhile, because it has angular momentum L, the Earth must also be moving around the Sun. In other words the angle θ is increasing with time. The resulting trajectory, in which the distance oscillates and the angular position changes, is elliptical. Figure 5 shows just such an elliptical orbit. If you follow the orbit and keep track only of the radial distance, the position of the Earth periodically moves in and out as if it were oscillating in the effective potential.

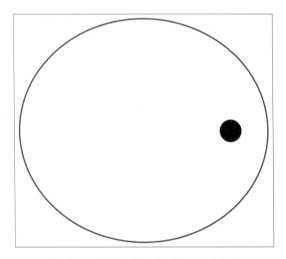

Figure 5: The elliptical orbit of the Earth around the Sun.

To prove the orbit is exactly an ellipse is a bit difficult, and we will not prove it now.

Let's take another look at the motion of a particle in the

effective potential. Imagine a particle with so much energy that it would completely escape from the dip in the potential energy. In such an orbit the particle comes in from infinity, bounces off the potential near $r = 0$, and goes back out, never to return. Such orbits certainly exist; they are called unbounded hyperbolic orbits.

Now let's move on to K2. According to Kepler's second law, as the radial vector sweeps out the ellipse, the area that it sweeps per unit time is always the same. This sounds like a conservation law, and indeed it is—the conservation of angular momentum. Go back to Eq. (7) and divide it by the mass m:

$$r^2 \dot{\theta} = \frac{p_\theta}{m}. \tag{15}$$

Imagine the radial line as it sweeps out an area. In a small time δt, the area changes by $\delta \theta$.

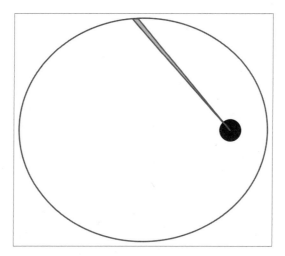

Figure 6: The area swept out by the line connecting the Earth to the Sun in a short time δt.

The small triangle swept out in Figure 6 has area

$$\delta A = \frac{1}{2} r^2 \, \delta \theta.$$

You can check this using the fact that the area of a triangle is one-half the base (r) times the height ($r\,\delta\theta$). If we divide by the small time interval δt, we get

$$\frac{dA}{dt} = \frac{r^2}{2} \dot{\theta}.$$

But now we use angular momentum conservation in the form of Eq. (15), and we get the final equation

$$\frac{dA}{dt} = \frac{p_\theta}{2m}. \tag{16}$$

Since p_θ (and also m) do not vary, we see that the rate of swept-out area is constant, and, moreover, it is just proportional to the angular momentum of the orbit.

Finally, we come to K3: *The square of the orbital period of a planet is directly proportional to the cube of the radius of its orbit.*

Kepler's formulation was very general, but we will work it out only for circular orbits. There are a number of ways we can do this, but the simplest is just to use Newton's law, $F = m\,a$. The force on the orbiting Earth is just the gravitational force, whose magnitude is

$$F = -\frac{GMm}{r^2}.$$

On the other hand, in Lecture 2 we calculated the acceleration of an object moving in a circular orbit,

$$a = \omega^2 r \tag{17}$$

where ω is the angular velocity.

Exercise 1: Show that Eq. (17), above, is a consequence of Equations (3) from Lecture 2.

Newton's law becomes

$$\frac{GMm}{r^2} = m\omega^2 r.$$

We can easily solve this for ω^2:

$$\omega^2 = \frac{GM}{r^3}.$$

The last step is to note that the period of the orbit—the time to make one circuit—is simply related to the angular velocity. Denoting the period by the Greek letter tau, τ, we have

$$\tau = \frac{1}{2\pi\omega}.$$

Traditionally we would use T for the period, but we are already using T for the kinetic energy. Putting it all together we get

$$\tau^2 = \frac{1}{4\pi GM} r^3.$$

Indeed, the square of the period is proportional to the cube of the radius.

Index